Inside the Stars

Online at: https://doi.org/10.1088/2514-3433/acce33

AAS Editor in Chief

Ethan Vishniac, Johns Hopkins University, Maryland, USA

About the program:

AAS-IOP Astronomy ebooks is the official book program of the American Astronomical Society (AAS) and aims to share in depth the most fascinating areas of astronomy, astrophysics, solar physics, and planetary science. The program includes publications in the following topics:

GALAXIES AND COSMOLOGY

INTERSTELLAR MATTER AND THE LOCAL UNIVERSE

STARS AND STELLAR PHYSICS

EDUCATION, OUTREACH, AND HERITAGE

HIGH-ENERGY PHENOMENA AND FUNDAMENTAL PHYSICS

THE SUN AND THE HELIOSPHERE

THE SOLAR SYSTEM, EXOPLANETS, AND ASTROBIOLOGY

LABORATORY ASTROPHYSICS, INSTRUMENTATION, SOFTWARE, AND DATA

Books in the program range in level from short introductory texts on fast-moving areas, graduate and upper-level undergraduate textbooks, research monographs, and practical handbooks.

For a complete list of published and forthcoming titles, please visit iopscience.org/books/aas.

About the American Astronomical Society

The American Astronomical Society (aas.org), established 1899, is the major organization of professional astronomers in North America. The membership (~7,000) also includes physicists, mathematicians, geologists, engineers, and others whose research interests lie within the broad spectrum of subjects now comprising the contemporary astronomical sciences. The mission of the Society is to enhance and share humanity's scientific understanding of the universe.

Inside the Stars

Hugh Van Horn

Professor of Physics and Astronomy Emeritus, University of Rochester, New York, USA

IOP Publishing, Bristol, UK

ISBN   978-0-7503-5794-4 (ebook)
ISBN   978-0-7503-5792-0 (print)
ISBN   978-0-7503-5795-1 (myPrint)
ISBN   978-0-7503-5793-7 (mobi)

DOI   10.1088/2514-3433/acce33

Version: 20230801

AAS–IOP Astronomy
ISSN 2514-3433 (online)
ISSN 2515-141X (print)

British Library Cataloguing-in-Publication Data: A catalogue record for this book is available from the British Library.

Published by IOP Publishing, wholly owned by The Institute of Physics, London

IOP Publishing, No.2 The Distillery, Glassfields, Avon Street, Bristol, BS2 0GR, UK

US Office: IOP Publishing, Inc., 190 North Independence Mall West, Suite 601, Philadelphia, PA 19106, USA

This book is dedicated to all the scientists—both women and men—who have been, are now, or in the future will be working to increase our understanding of the Universe in which we live.

Contents

Preface

The growth in our understanding of the interiors of stars is a fascinating story that dates back to the last half of the 19th century. No one then had any idea what the stars are made of, what determines their internal structure, or what the sources of energy are that produce their power output, even for the closest and most well-studied star, our Sun. Most scientists then thought that the Sun was some kind of hot, solid body, like a larger and much hotter version of a planet. Today, in contrast, we speak authoritatively about the internal properties of many different kinds of stars. But how do we *know*? The purpose of this book is to explain how astronomers and astrophysicists determined the internal properties of the stars and to describe how revolutionary advances in observational astronomy and physics, coupled with astonishing advances in technology, instrumentation, and computing power, have enabled these discoveries.

The enormous temperatures and densities we now know to be present inside the stars cannot be duplicated in terrestrial laboratories. Consequently, scientists who wish to understand the properties of matter under these conditions have no choice but to resort to theory. In the physical sciences, "theory" usually means doing some kind of calculation, either using a system of equations to compute with mathematical precision a result that can in principle be compared with some type of measurement in the physical world, or else working to develop new systems of equations that fit existing data better than previous ones. The computation of spacecraft trajectories is an example of the former, while Kepler's discovery that planetary orbits are ellipses—rather than circles with added epicycles—is an example of the latter.

In contrast to theory, experiment and observation involve direct measurements of the physical world. Experiments involve physical systems that can be subjected to strict control—such as measurements made at specific temperatures, pressures, or magnetic field strengths in a terrestrial laboratory—while observations involve systems that are beyond human control. Astronomical observations of any kind are examples of the latter.

The purpose of this book is to tell the story of the way astronomers and physicists gradually combined forces to achieve an increasingly detailed understanding of the Universe, creating the discipline we now call astrophysics.

Hugh M. Van Horn
Marlton, NJ

Acknowledgments

I wish to express my gratitude to the late Edwin E. Salpeter, who first introduced me to the physical properties of the dense matter inside stars, and to the late Malcolm P. Savedoff, who aroused my interest in white dwarf stars, objects with masses comparable to that of the Sun crammed into dimensions about the sizes of planets. I am also grateful to my former students and postdocs, who shared with me the excitement of discovering new information about the interiors of the stars. In addition, I thank my astrophysics colleagues around the world for allowing me to share in this journey of discovery. Last but by no means least, I am deeply grateful to my wife, Sue, for her patience while I spent time working on this book and for reading and commenting on an earlier draft.

Author biography

Hugh M. Van Horn

 **Hugh M. Van Horn** was a member of the University of Rochester's Department of Physics and Astronomy for nearly three decades, becoming Professor Emeritus of Physics and Astronomy in 2005. He joined the National Science Foundation in Washington, DC, in 1993, where he served as Director of the Division of Astronomical Sciences and later as Director of National Facilities for the Division of Materials Research. After retiring in 2004, he has devoted his time to working as an author and as a freelance editor. His previous astronomy book, entitled *Unlocking the Secrets of White Dwarf Stars*, was published by Springer in 2015.

Part I

Understanding the Inside of the Sun

Our understanding of the interiors of stars began late in the 19th century when scientists first began to learn about the inside of our Sun, the closest star. The first part of this book tells the story of those first steps and the ones that followed over the next several decades.

AAS | IOP Astronomy

Inside the Stars

Hugh Van Horn

Chapter 1

A Glimpse of the Solar Interior

At the beginning of the 19th century, even the most basic information about the Sun was known poorly, if at all. For instance, how far away is it from Earth? This distance—known today as the astronomical unit, or AU—is one of the fundamental quantities needed to determine the radius, mass, and luminosity (total power output) of the Sun. It is also essential for determining the distances to other stars and for measuring their masses and luminosities.

Recognizing its importance, astronomers have endeavored since antiquity to determine the Earth–Sun distance using a variety of different methods. By the early 19th century, Johann Franz Encke (1791–1865) was able to combine all the available data to obtain a value for the astronomical unit that was accurate to better than half a percent (Pannekoek 1961a). He found one AU to be equal to 153.3 million km; for comparison, the modern value for this quantity, obtained from radar measurements, is 1 AU = 149.59787066 million km.[1]

From the known angular diameter of the Sun—a bit more than half a degree of arc—this distance immediately gives the solar radius as $R_{\mathrm{Sun}} = 714{,}000$ km from high-school trigonometry. Again, the modern value is similar: $R_{\mathrm{Sun}} = 695{,}508$ km.

The value of the AU also enabled astronomers to determine the solar mass. In 1619, Johannes Kepler (1571–1630) published his third law of planetary motion, showing that the square of the period of a planetary orbit around the Sun is proportional to the cube of half of the major axis (called the "semi-major axis") of the elliptical orbit (Abell et al. 1987a). In 1687, Sir Isaac Newton (1643–1727) explained this result as a consequence of his laws of motion and his theory of universal gravitation, and he found that the constant of proportionality involves the sum of the masses of the planet and the Sun. Since the semi-major axis of Earth's orbit about the Sun is approximately equal to one AU, since we know the period of Earth's orbit (one year), and since the mass of the Earth is much smaller than the

---

[1] This and other modern values for astronomical quantities quoted in this chapter are from Cox (2000, p. 12).

doi:10.1088/2514-3433/acce33ch1

mass of the Sun, it was immediately possible to determine the solar mass. The result was $M_{Sun} = 2 \times 10^{30}$ kilograms; the modern value is $M_{Sun} = 1.9891 \times 10^{30}$ kg.[2]

The determination of the total power output of the Sun—called its "luminosity," L_{Sun}—required an additional measurement: the amount of energy per second from sunlight falling on a unit area of Earth's surface—a quantity called the "solar constant." French physicist Claude Pouillet (1790–1868) made the first measurement of this quantity in 1838, obtaining a value of slightly more than 1.2 kilowatts per square meter (kW·m^{-2}) (http://en.wikipedia.org/wiki/Solar_constant#Historical_measurements). Since every square meter of an imaginary sphere with a radius equal to that of Earth's orbit (approximately 1 AU) and concentric with the Sun receives the same amount of solar energy, we can find the total power output of the Sun by multiplying the "solar constant" by the surface area of that sphere. The result is $L_{Sun} = 3.5 \times 10^{23}$ kilowatts; the modern value for this quantity is $L_{Sun} = 3.8458 \times 10^{23}$ kilowatts.

Another important 19th-century advance was the discovery of the elemental compositions of the surfaces of the Sun and the stars. This developed from Joseph Fraunhofer's (1781–1826) serendipitous discovery of narrow, dark lines in the solar spectrum (see Figure 1.1), which he published in 1817 (Pannekoek 1961b). No one knew what to make of this until the German physicist Gustav Kirchoff (1824–1887) put the understanding of emission and absorption of radiation by solids, liquids, and gases on a solid foundation in 1859. His three laws of radiation (Abell et al. 1987b) state:

1. "A luminous solid or liquid emits light at all wavelengths, thus producing a continuous spectrum."
2. "A rarified luminous gas emits light whose spectrum shows bright lines..."
3. "If the white light from a luminous source is passed through a gas, the gas may abstract certain wavelengths from the continuous spectrum..., thus producing dark lines."

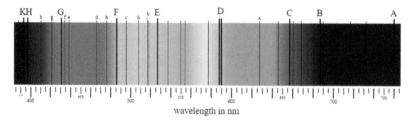

Figure 1.1. Fraunhofer lines in the solar spectrum. The horizontal axis is the wavelength in nanometers (nm); one nanometer (10^{-9} meters) equals 10 Ångstroms. Fraunhofer labeled the strongest lines with capital letters. Source: https://commons.wikimedia.org/wiki/File:Fraunhofer-lines.jpg.

---

[2] For simplicity, astronomers use powers of ten to represent very large numbers. For example, $100 = 10^2$, $1000 = 10^3$, 1 million = $1,000,000 = 10^6$, and so on. Thus, 10^{30} kg = $10^6 \times 10^6 \times 10^6 \times 10^6 \times 10^6$ kg = one million million million million million kilograms.

During the next few years, Kirchoff showed how this new understanding could be used to analyze the spectra of the Sun and stars to determine their chemical makeup. Concentrating on the Sun, he "measured ... the positions of some thousands of Fraunhofer lines and established their coincidence with lines emitted by diverse chemical elements such as hydrogen, iron, sodium, magnesium, calcium, etc. He concluded that these elements were present in the Sun's atmosphere, absorbing [at] their special wavelengths from the continuous spectrum emitted by the solar body" (Pannekoek 1961c).

By the end of the 19th century, scientists had thus discovered that the Fraunhofer lines labeled H and K in Figure 1.1 were produced by Ca ions, and G turned out to be produced by a rotation–vibration band of the CH molecule. The line labeled F was the second in order of decreasing wavelength—today called the "Hβ" line—in the series of neutral-hydrogen absorption lines discovered in 1885 by Swiss scientist Johann Jakob Balmer (1825–1898) (http://en.wikipedia.org/wiki/Johann_Jakob_ Balmer). The strong line Fraunhofer labeled E was actually a group of Fe absorption lines, while the doublet labeled D was produced by neutral sodium; in emission—as indicated by Kirchoff's second law—it produces the familiar yellow color of sodium-vapor lights. Line C was the Hα line in the Balmer series, and Fraunhofer lines A and B turned out to be produced by O_2 molecules in Earth's atmosphere.

> "The new discoveries brought new ideas on the nature of the Sun. It was no longer possible to believe that the Sun's interior was a dark and cold body. Kirchoff considered the Sun to be a red-hot sphere, solid or liquid ... surrounded by a less hot atmosphere containing the terrestrial elements in a gaseous state, which produce the Fraunhofer lines..." (http://en.wikipedia.org/wiki/Johann_Jakob_Balmer).

The first detailed theoretical calculation for any star was J. Homer Lane's 1870 publication of a mathematical model for the internal structure of the Sun (Lane 1870a). By all accounts, Jonathan Homer Lane (1819–1880; Figure 1.2) was a remarkable individual.[3] He was born in 1819 in the village of Genesee—today called Geneseo—in western New York State. Located just to the west of the picturesque Finger Lakes district, the town is situated amid fertile farmlands along the flood-plain of the Genesee River. Homer's father was a farmer, and Homer and his brother and two sisters no doubt helped with chores from the time they were old enough to do so. Formal education was not a priority for farm children, and Homer consequently attended school only until he was eight years old. Thereafter he was home-schooled by his parents.

---

[3] Much of the material in this chapter is taken from Abbe, C. 1895, BMNAS, **3**, 253, "Memoir of Jonathan Homer Lane, 1819–1880" and from Powell, C. 1988, JHA, **19**, 183, "J. Homer Lane and the Internal Structure of the Sun."

J. HOMER LANE

AUGUST 9. 1819— MAY 3. 1880.

Figure 1.2. J. Homer Lane. This pencil sketch was "made from life by his friend, Dr. E. W. Schaefer, of Washington, [DC, and is] dated May 16, 1868…" (See 1906). It is the only known image of Lane in existence. Source: http://www.astro.yale.edu/demarque/lane.html; used with permission from P. Demarque.

Lane evidently had a gift for science and mathematics, however, as he was admitted to the Phillips Academy in Exeter, New Hampshire, at age 20. He subsequently enrolled at Yale, and personal records from this period reveal an early interest in astronomy (Powell 1988a). He also developed the habit of reading the major scientific periodicals of the day to keep abreast of the latest scientific discoveries and advances (Powell 1988b). In addition, he began to correspond with the celebrated American physicist Joseph Henry (1797–1878) during his college years (Powell 1988a), and they maintained regular communications for the rest of their lives.

Lane graduated from Yale in 1846, and the first of his publications appeared that year. The Smithsonian Institution in Washington, DC, was founded in the same year, with Professor Henry selected as the first Secretary. Henry had clearly been impressed with Lane's abilities, as he recommended the young man for an appointment as Associate Examiner in the United States Patent Office. Lane was evidently good at his job, because he was promoted to Principal Examiner three years later.

During his tenure as a patent examiner, Lane was described as "laborious and thorough, cautious and critical, conscientious in the extreme" (Abbe 1895a). At the same time, "Professor Henry ... took great pleasure in ... drawing out from the reserved and rather silent student his views on abstruse or doubtful physical questions" (Abbe 1895a). And both Henry and Lane were among the initial "members of a scientific and social club organized at Washington [about 1857, which] ... in 1871 ... developed into the Philosophical Society of Washington" (Abbe 1895a).

Lane's interest in astronomy was piqued during this period in the Patent Office. In 1854, the noted German scientist Hermann von Helmholtz (1821–1894) delivered a popular lecture in Königsberg in which he postulated that the energy radiated by the Sun might be due to the heat produced by a slow contraction under gravity (Chandrasekhar 1939a). Lane thus began to wonder "Whether the entire mass of the Sun might not be a mixture of ... gases ..." (Lane 1870a) and he started "thinking about the balance between gravity and gas pressure in the Sun's interior..." (Powell 1988c). At the time, "the constitution of the Sun was exciting ... much discussion..." (Lane 1870b).

Lane's life was uprooted in the late 1850s, however, when he was unceremoniously removed from his position at the Patent Office to make room for a political appointee. For the next three years he maintained an office in Washington, where he operated a small business as an expert and counselor in patent cases. With the opening of oil fields in Pennsylvania, however, Lane saw an opportunity to make money speculating in the sale of oil lands. He accordingly moved to Franklin, Pennsylvania, in 1860, where he lived with his brother (Powell 1988d).

During the six years he spent in Pennsylvania, Lane began to think about the internal structure of the Sun in thermodynamic terms (Powell 1988e). During the 1840s and 1850s, physicists were developing an understanding of heat as a form of energy, establishing the absolute scale of temperature, and developing the theory of thermodynamics, which relates changes in the heat content of a body to changes in other physical properties.

Lord Kelvin's groundbreaking work on convective equilibrium, which appeared in 1862 (Powell 1988f; Chandrasekhar 1939b), was one of three key concepts that enabled Lane to develop a mathematical model for the Sun. The other two—the concept of the Sun as a wholly gaseous body and the idea that heat could be transported through the solar interior by convective motions—appear to have originated with Lane himself (Powell 1988f). Lane kept detailed private notes of his ideas in "... a small collection of unpublished papers entitled 'The Sun viewed as a gaseous body'..." (Powell 1988e).

When Lane returned to Washington in 1866, he had accumulated significant wealth from the sale of oil lands. After a few more years, supporting himself—and aiding his brother and sisters—from these funds, he "accepted a position under the Superintendent of the Coast Survey as a 'Verifier of Standards in the Office of Weights and Measures' ..." (Abbe 1895b) which he retained for the rest of his life.

According to Julius E. Hilgard, Superintendent of the U.S. Geodetic and Coast Survey, "The quality of Mr. Lane's mind was truly remarkable, being chiefly characterized by an extraordinary precision of thought and logic ... [but unfortunately] he lacked fluency of speech. Of the quietest, most retiring disposition, he was ... known to but few; diffident in manner and not given to many words, yet when he did speak his rare logic was such as to carry conviction. In his writings his clearness of mind became manifest ..." (Abbe 1895b).

Shortly after Lane's return to Washington, his "associate Benjamin Pierce gave a lecture entitled 'On the origin of solar heat' ... [concluding] that Helmholtz's gravitational contraction hypothesis provided a plausible heat source for the Sun..." (Powell 1988f). Lane was also prompted to develop his ideas further by communications by his friend Dr. Craig and astronomer Simon Newcomb "to a company of scientific gentlemen early last spring" (Lane 1870b). Presumably this was the spring of 1868.

Lane approached the problem not as an astronomer but as a mathematical physicist, in that he proposed a physical model—a wholly gaseous Sun in hydrostatic equilibrium, through which heat is transported by convection—and then worked out the consequences mathematically (Powell 1988g). He thus avoided many misconceptions about the Sun that were prevalent at the time. Indeed, even as late as 1872, one authority wrote that "certain portions of the solar globe ... [probably exist] as liquids or even as solids" (Powell 1988g). Lane's hypothesis that the Sun be considered as a wholly gaseous body was a departure from then-current thinking in astronomy, and there was no existing evidence to support it.

Reasoning that the observed stability of the Sun over time required that it be in hydrostatic equilibrium, Lane was able to write down a differential equation[4] that expressed the pressure gradient at any spherical shell within the Sun in terms of the gravitational force exerted by the mass inside the shell. That mass, in turn, is determined by the distribution of the mass density. And if the Sun were an ideal gas, as Lane postulated, another well-known equation related the pressure to the local density and temperature. All that was needed in order to enable Lane to solve his set of equations was an expression relating the temperature to the pressure and density. This was provided by Lord Kelvin's concept of convective equilibrium.

Lane's solution of the equations he had developed to describe the internal structure of a wholly gaseous Sun yielded the distribution of mass density as a function of distance from the solar center as well as the distribution of temperature relative to its central value. He published graphs of these functions for each of the two cases he had calculated. Today, we would classify these as polytropes of index $n = 1.5$ and index $n = 2.5$.[5] Lane's graph of the relative distribution of temperature for these two cases (Lane 1870c) is shown in Figure 1.3, together with a similar curve for the relative temperature distribution in the Sun obtained from a detailed numerical model computed in 1957 (Weymann 1957).

---

[4] A differential equation is one that relates infinitesimal changes in one variable—such as distance traveled—to infinitesimal differences in some quantity—such as time—over which the variable changes.

[5] A polytrope of index n is a mathematical model in which the pressure is proportional solely to a power of the density, with the exponent of this power law written as $(n+1)/n$.

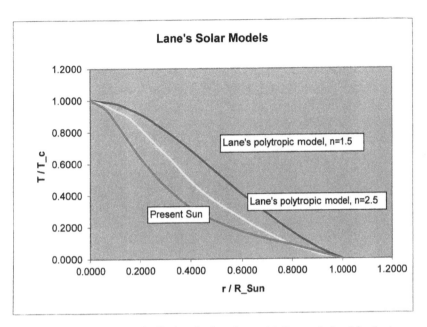

Figure 1.3. The relative temperature distributions in the solar models Lane calculated for the two cases with polytropic indices $n = 1.5$ (the blue curve) and $n = 2.5$ (the yellow curve). The data are from Lane (1870), p. 69, as plotted by the present author. For comparison, the red curve shows a model for the present Sun computed by Ray Weymann (Weymann 1957) reproduced in Schwarzschild (1958), p. 259. Here T_c is the central temperature of the Sun, and R_{Sun} is the solar radius. The quantity T is the local temperature at radius r.

Of course, the surface temperature of a real star is not zero; that is an artifact of the simplified nature of the polytropic models. Because the density and pressure at the surface of a real star are so much smaller than the central density and pressure— they approach the vanishingly small density and pressure of space—it is reasonable to approximate them as zero at the surface of a polytropic model. And because the temperature in a polytrope is proportional to a power of the density (or pressure), the surface temperature in such a model is forced to become vanishingly small as well. This is also not unreasonable; the effective temperature[6] of the Sun, about 5800 K, is more than two thousand times smaller than its central temperature.

In addition, Lane calculated explicitly the central densities for these two models, obtaining 7.11 g cm^{-3} for the model with index $n = 1.5$ and 28.16 for $n = 2.5$. The latter, he noted, was "nearly one-third greater than that of the metal platinum" (Lane 1870d).

---

[6] The effective temperature of a star is the temperature of an ideal blackbody with the same radius and luminosity. It can be thought of as being approximately the surface temperature of the star. A "blackbody" absorbs all radiation incident upon it and converts it into internal heat, which it reradiates as a continuous spectrum that depends only upon the temperature of the blackbody. The Kelvin scale of temperature is the same as the Celsius (or centigrade) temperature scale except that the zero of temperature is at absolute zero, where all thermal motions cease, approximately −273 °C.

Lane's mathematical model for the Sun demonstrated for the first time that our nearest star can be understood on the basis of physical laws and properties of matter as determined in terrestrial laboratories. Of course, Lane's model was still very approximate. At the time of his work, no one knew the source of the solar energy. And quantum theory, which was to enable calculations of the material properties and of the interaction between matter and radiation—both central to the understanding of the internal structures of the stars—still lay a half century in the future.

As pointed out by the brilliant Indian astrophysicist Subrahmanyan Chandrasekhar (1910–1995) in his classic text on stellar structure, Lane's 1870 paper "made him the author ... of the first investigation on stellar interiors...." (Chandrasekhar 1939c), adding that "When the crudeness of the then available data is considered, Lane's success in estimating density and temperature at the center of the Sun is a remarkable achievement" (Chandrasekhar 1939d). According to Corey Stevenson Powell, "Perhaps as a result of his paper, Lane was elected a member of the National Academy of Sciences in April 1872..." (Powell 1988h).

Following Lane's inspired work, the focus of effort in understanding stellar structure moved to Europe. In the five-year period 1878–1883, Georg August Dietrich Ritter (1826–1908) developed mathematical models similar to Lane's, though Ritter's work was done completely independently and with no knowledge of Lane's efforts. Ritter's 18 highly technical papers, published in *Wiedemann's Annalen*, developed the mathematical theory of polytropes in much greater detail than Lane had done in his one paper on the subject. According to Chandrasekhar, "Ritter's investigations are very remarkable in their range and depth..." (Chandrasekhar 1939d). In particular, in an 1880 paper, "Ritter established (explicitly for the first time) the fundamental ... equation governing the structure of gaseous configurations with an underlying [polytropic] law .... [Almost] the entire foundation for the mathematical theory of stellar structure was laid by him" (Chandrasekhar 1939d). And in 1907, Swiss astrophysicist Jacob Robert Emden (1862–1940) published his classic work, *Gaskugeln* (*Gas Spheres*, in English), which summarized all that was then known—including the results of his own researches—about (celestial) spheres supported by gas pressure against their own self-gravity. "The publication of Emden's *Gaskugeln* marks the end of the first epoch in the study of stellar configurations" (Chandrasekhar 1939e). In his classic text on stellar astrophysics, Chandrasekhar devoted an extensive chapter to the mathematical theory of polytropes. He summarized the earlier works by Lane, Ritter, Emden, and Lord Kelvin, and tabulated a number of key properties of the models (Chandrasekhar 1939f) that enable comparisons to the physical properties of the Sun and the stars. These include the model's radius, mass, the mass–radius relation, the ratio of the central to the mean density, and the central pressure.

So, what did these early investigations reveal about the internal structure of the Sun?

First, Lane's polytropic model for a gas sphere with $n = 1.5$ and the mass and radius of the Sun as then known had a central density of 7.11 grams per cubic centimeter (g/c.c. or g cm^{-3}). Modern values for the solar mass and radius and the more accurate polytropic constants tabulated by Chandrasekhar (1939f) give a density of 8.46 g cm^{-3}, about the density of solid copper, and a central pressure of 8.69 × 10^{15} dynes per square centimeter (dynes·cm^{-2}). That pressure is equal to about 9 billion bars, or about 9 million times larger than the pressure at the bottom of the deepest ocean trenches on Earth.

Second, although Lane himself did not compute the central temperatures of his polytropic models, "Henry Norris Russell later noted that 'Since the work of Lane in 1870, it has been realized that, if the familiar laws of perfect gases could be applied to the interior of the Sun, the central temperature must be many millions of degrees'" (Russell 1939). For a pure hydrogen model with the central densities and pressures of Lane's two models, the central temperatures are 12.4 million degrees Kelvin[7] (for $n = 1.5$) and 16.1 million degrees Kelvin (for $n = 2.5$). These results—far higher than the temperatures of the hottest flames that can be achieved in a terrestrial laboratory, about 3800 K for an oxyacetylene torch—was astonishing to scientists of the day.

To make further progress in understanding the true nature of solar and stellar interiors, however, it was necessary to move beyond models based on such simplified approximations for the physical properties of stellar matter. Lane himself recognized this, writing in his 1870 paper that "it is of course easy enough to conceive that in the fierce collisions of ... molecules with each other at the temperatures supposed to exist in the Sun's body, their component atoms might be torn asunder and might thenceforth move as free simple molecules ... and in fact that all the matter in the Sun's body is split into simple free atoms..." (Lane 1870e). And in his 1909 review of Emden's classic treatise, Sir James H. Jeans wrote, "Before astrophysical truths can be derived by mathematical analysis it appears that new terms must be introduced into the ordinary gas-equations to represent light-pressure, electric action, and possible cohesion and other agencies" (Jeans 1909).

There were good reasons to be concerned, because during the final decade of the 19th century and the first decades of the 20th, our understanding of the physical properties of matter and light were undergoing a profound transformation.

References

Abbe, C. 1895a, BMNAS, 3, 256

Abbe, C. 1895b, BMNAS, 3, 259

Abell, G. O., Morrison, D., & Wolff, S. C. 1987a, Exploration of the Universe (Philadelphia, PA: Saunders College Publishing) 39

Abell, G. O., Morrison, D., & Wolff, S. C. 1987b, Exploration of the Universe (Philadelphia, PA: Saunders College Publishing) 149

Chandrasekhar, S. 1939a, An Introduction to the Study of Stellar Structure (Chicago, IL: Univ. Chicago Press) 454

---

[7] The absolute scale of temperature was defined by Lord Kelvin in 1848 and bears his name, symbolized by the letter "K." As noted previously, it is similar to the centigrade or Celsius scale of temperature, except that zero Kelvin (called "absolute zero") is the temperature at which all thermal motions vanish. In contrast, zero degrees centigrade is the temperature at which water freezes (equal to 32 °F), while 100 °C is the temperature at which water boils (212 °F).

Chandrasekhar, S. 1939b, An Introduction to the Study of Stellar Structure (Chicago, IL: Univ. Chicago Press) 84

Chandrasekhar, S. 1939c, An Introduction to the Study of Stellar Structure (Chicago, IL: Univ. Chicago Press) 176

Chandrasekhar, S. 1939d, An Introduction to the Study of Stellar Structure (Chicago, IL: Univ. Chicago Press) 177

Chandrasekhar, S. 1939e, An Introduction to the Study of Stellar Structure (Chicago, IL: Univ. Chicago Press) 180

Chandrasekhar, S. 1939f, An Introduction to the Study of Stellar Structure (Chicago, IL: Univ. Chicago Press) 96

http://en.wikipedia.org/wiki/Johann_Jakob_Balmer

http://en.wikipedia.org/wiki/Solar_constant#Historical_measurements

Jeans, J. H. 1909, ApJ, 30, 72

Lane, J. H. 1870a, AmJS, 2nd ser., 50, 57

Lane, J. H. 1870b, AmJS, 2nd ser., 50, 58

Lane, J. H. 1870c, AmJS, 2nd ser., 50, 69

Lane, J. H. 1870d, AmJS, 2nd ser., 50, 63

Lane, J. H. 1870e, AmJS, 2nd ser., 50, 71

Pannekoek, A. 1961a, A History of Astronomy (London: George Allen & Unwin Ltd.) 341

Pannekoek, A. 1961b, A History of Astronomy (London: George Allen & Unwin Ltd.) 330

Pannekoek, A. 1961c, A History of Astronomy (London: George Allen & Unwin Ltd.) 407

Powell, C. 1988a, JHA, 19, 184

Powell, C. 1988b, JHA, 19, 188

Powell, C. 1988c, JHA, 19, 187

Powell, C. 1988d, JHA, 19, 185

Powell, C. 1988e, JHA, 19, 186

Powell, C. 1988f, JHA, 19, 189

Powell, C. 1988g, JHA, 19, 190

Powell, C. 1988h, JHA, 19, 193

Russell, H. N. 1939, PAPhS, 81, 295
    quoted in Chandrasekhar, S. 1939, An Introduction to the Study of Stellar Structure (Chicago, IL: Univ. Chicago Press) 192

See, T. J. J. 1906, PA, 14, 193

Weymann, R. 1957, ApJ, 126, 208
    summarized in Schwarzschild, M. 1958, Structure and Evolution of the Stars (Princeton, NJ: Princeton Univ. Press) 259

Chapter 2

The Sun as a Star

The idea that stars might be similar to the Sun—except for being located at much greater distances—seems to have originated at least hundreds of years ago. However, not until the 19th century were astronomical instruments and observing techniques sufficiently accurate to enable astronomers to make the precise measurements necessary to determine the distances to the stars. In 1838, the famed German astronomer and mathematician Friedrich W. Bessel (1784–1846) became the first to accomplish this, when he measured the parallax of the binary star system 61 Cygni. The process he used to measure astronomical distances would be familiar in principle to any Earth-bound surveyor: you first measure accurately the angular bearing from your present location to the object whose distance you want to determine. Then move a known distance to another location and measure the angular bearing to the point of interest from there. From the two measured angles and the known distance between the two locations from which you took the bearings, you can determine the desired distance, just as you learned in high-school trigonometry class.

How does this work in astronomy, though? After all, unless you happen to be in an interplanetary spacecraft, you can't move from one point in space to another point a known distance away. The answer, of course, is that we are all traveling together through the solar system on our very own spacecraft: the planet Earth.

Suppose we observe a relatively nearby star against the background of more distant stars from some given point in Earth's orbit around the Sun. That is, we measure as precisely as we can the exact locations of a large number of stars on the celestial sphere.[1] We can determine each star's location by measuring its celestial

---

[1] The celestial sphere is an imaginary sphere centered on the Earth on which we can plot the positions of celestial objects using celestial angular coordinates like the latitude and longitude used to fix the locations of points on the Earth's surface.

longitude and latitude—quantities that astronomers call its "right ascension" and "declination," respectively.

Now wait six months (while you are of course doing something else). During this time, the Earth moves halfway around its orbit about the Sun. Since the average distance from the Earth to the Sun is by definition one astronomical unit (AU), the Earth's new position is two AU away from its original point in the orbit. This gives the known distance for our surveyor's measurement as two AU.

Now measure the celestial coordinates of all the stars in your original sample again as accurately as possible. The coordinates for most of the stars will be the same as they were six months earlier, to within the accuracy of the measurement. But some—the nearer stars—will be found to have slightly different coordinate values. This is the result of parallax: nearer objects appear to move against a more distant background when the observing location changes.

You can easily see this effect for yourself. For example, suppose you are looking at a tree standing in a field, with a range of hills in the background. Close one eye and observe carefully the position of the tree against some landmark on the hills. Now open that eye and close the other. The tree will appear to have moved slightly against the distant hills because the viewpoint of your left eye is separated by several inches from the viewpoint of your right eye. That is the parallax effect.

Returning to the stars, Bessel used this slight shift in the apparent position of the nearer stars against the background of the more distant ones to determine the parallax angle. From the parallax angle and the known baseline distance, he was then able to determine the distances to the nearest stars (see Figure 2.1). For the visual binary 61 Cygni, he found the tiny parallax angle to be 0.31 seconds of arc (or, 0.31″). That value is roughly equivalent to measuring the size of a nickel at a distance of 13 km. Re-measuring the parallax two years later, Bessell obtained the more accurate value 0.348″, corresponding to a distance of 590,000 AU (Pannekoek 1961a). Astronomers were astonished to find that even the nearest stars lie at enormous distances beyond the solar system!

Rather than measuring such large distances in terms of AU or kilometers, astronomers find it more convenient to use as a unit the distance at which a star would have a parallax of one second of arc, a distance they term one "parsec" (abbreviated "pc").[2] And a star 10 parsecs away has a parallax of one-tenth of a second of arc. Since 60 seconds of arc equal one minute of arc, and 60 minutes of arc equal one degree, one arcsecond is 1/3600 of a degree, or about 1800 times smaller than the angular diameter of the full Moon.

Determinations of stellar distances also enabled astronomers to ascertain other properties of the stars. For example, just as for the case of the Sun, stellar masses were obtained by applying Newton's generalization of Kepler's third law—in the stellar case, to binary systems. This approach gives the sum of the masses of the two stars (in units of the solar mass, M_{Sun}) as the cube of the sum of the semi-major axes of the orbits of the two stars (in AU) about their common center of mass, divided by

---

[2] One parsec equals 3.26 light-years—the distance light travels in one year—or 3.09×10^{13} km.

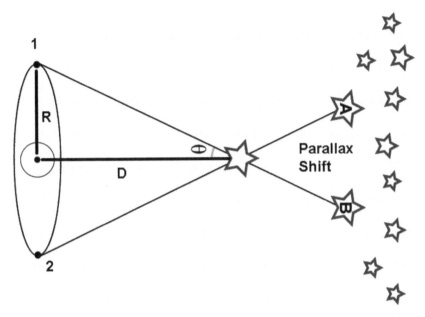

Figure 2.1. Determining stellar distances. The ellipse at the left represents Earth's orbit around the Sun. The distance from the Earth to the Sun, represented by R in the figure, is one astronomical unit (AU). If an observation of a star located a distance D away is made at the point labeled "1" in Earth's orbit, the star appears at the position labeled "B" against the more distant background stars. Six months later, when the Earth is at position "2" in its orbit, another observation of the target star shows it to appear at position "A." The apparent shift is called the parallax of the star, and the distance D can be determined from trigonometry using the measured parallax angle θ and the known Earth–Sun distance. Source: http://image.gsfc.nasa.gov/poetry/venus/Vdistance.html; image courtesy of NASA.

the square of the orbital period (in years). In order to use this method, the astronomer needs an accurate measurement of the period and a precise determination of the orbit. The distance to the binary enters this calculation because it is needed to convert the observed angular separations of the two stars into distances measured in AU. According to Pannekoek, "Herschel had published lists of hundreds of double stars and in 1803 found the relative motions of the components for about 50 of them … [but his] pioneer work …was too coarse for the demands of a good determination of orbits." Thus, double-star astronomy "became the first field of application for the refined 19th-century instruments, with their higher standard of accuracy" (Pannekoek 1961b). By 1850, some 20 orbits had been measured, and this kind of work continued throughout the rest of the century (Aitken 1918).

What were the results of these mass determinations? In a list compiled by Hertzsprung in 1923 (Hertzsprung 1923) the masses of the components of 15 visual binaries were found to range from 0.14 M_{Sun} to 4.18 M_{Sun} (Capella). The mass of Procyon is given in the list as 1.13 M_{Sun} and that of α Centauri as 1.14 M_{Sun}.[3] Fifteen

---

[3] For comparison, the modern values are 1.69 M_{Sun} for Procyon and 1.08 M_{Sun} for α Centauri: Cox 2000, p. 424.

of the 21 individual components are less massive than the Sun. Conversely, all but one of 13 eclipsing variables[4] and all five Cepheid variables[5] exceed the solar mass. So far as the masses are concerned, the Sun thus appears to be a pretty average star.

Because they are so far away, the stars appear to us as tiny points of light, even though they may be as large and bright as the Sun. In fact, many are much brighter, and many more are fainter. In the second century BCE, the Greek astronomer Hipparchus classified the brightest stars he could see as those of "first magnitude" and the faintest as "sixth magnitude." When quantitative measurements of the distances to the stars became possible, this qualitative classification system was converted into a quantitative one. By the mid-19th century, astronomers had found that a first-magnitude star was about 100 times brighter than a sixth-magnitude star. That is, the flux of energy received from a first-magnitude star is about 100 times greater than that from a sixth-magnitude star. Accordingly, in 1856 British astronomer Norman R. Pogson (1829–1891) proposed that astronomical magnitudes be redefined so that a difference of five magnitudes equals precisely a factor of 100 in brightness.[6] This recommendation was adopted and is still in use today (MacRobert 2006).

Astronomers also devised a way to compare the intrinsic brightnesses of the stars. At its actual distance, the brightness of a star is called its "apparent magnitude" and is designated by the lower-case symbol m. If the distance to the star is known, an astronomer can easily calculate the brightness it would have if transported—in his or her imagination—to a standard distance of 10 parsecs. The brightness of a star at this standard 10-parsec distance is termed its "absolute magnitude" and is designated by the upper-case symbol M. Throughout most of the 19th century, astronomers could only determine the apparent magnitudes of stars by eye; that is, by making comparisons with a standard source directly at the telescope. They were nevertheless able to obtain measurements of surprisingly good accuracy. When later methods of photometric measurement became possible—using photography in the late 19th and early 20th centuries and subsequently photoelectric photometry in the 20th century—it became necessary to distinguish between apparent magnitudes as measured with instruments having differing responses to the wavelengths of radiation. Then the apparent and absolute magnitudes measured with devices that approximated the visual response of the human eye were defined as apparent and absolute "visual" magnitudes, designated m_V and M_V, respectively.

As they determined absolute magnitudes for the stars, astronomers found that some of the more distant stars were intrinsically very much brighter than the stars

---

[4] An eclipsing variable is a binary star system in which the plane of the orbit lies in the line of sight from the Earth, so that as they revolve about each other in their orbit, first one star and then the other passes between the Earth and the other star in the binary system, eclipsing it.

[5] A Cepheid variable is a massive star that undergoes regular radial pulsations, with the stellar surface moving inward and outward with a regular period; see Chapter 20.

[6] Equivalently, the difference in magnitudes between a first- and a sixth-magnitude star is 5/2 times the logarithm of the relative energy fluxes.

that can be seen by eye. To accommodate these bright stars, astronomers extended the scale from first magnitude to zero and then on to negative values. The system of astronomical magnitudes thus has the peculiar feature—stemming from Hipparchus' original classification of stellar brightnesses millennia ago—that the very brightest stars have the largest negative magnitudes while the very faintest have the largest positive values, the reverse of what one might expect. Among the hundred nearest stars (Allen 1955a), Sirius is the brightest, with apparent visual magnitude −1.46, while one of the faintest is a very dim, cool, red companion to a star designated +4° 4048, with apparent visual magnitude +18.0.

The absolute magnitude proved to be directly proportional to the logarithm of the total power output of a star, a quantity astronomers term its "luminosity." The luminosity includes contributions from radiation at all wavelengths, not just in the visual band, and the corresponding absolute magnitude is termed the "absolute bolometric magnitude," designated M_{bol}, because at one time the total amount of radiation across all wavelengths was measured with a device called a bolometer. The relative amount of radiation at different wavelengths, compared to the amount in the visual wavelength band, depends upon the temperature of the star, however, so the difference between M_V and M_{bol}—a quantity called the "bolometric correction"—depends upon the star's temperature, too. How in heaven were astronomers going to be able to determine a star's temperature, though? After all, they could not go out to it and stick in a thermometer!

This problem was solved by discoveries in physics during the last two decades of the 19th century. In 1879, Austrian physicist Josef Stefan (1835–1893) showed from careful measurements at several different temperatures that the total power output of thermal radiation per unit area of an ideal emitting body[7] varies as the fourth power of the absolute temperature. Five years later, the noted German physicist Ludwig Boltzmann (1844–1906) provided a theoretical foundation for Stefan's experimental results. And in 1893, Wilhelm Wien (1864–1928) showed that the wavelength at which the spectrum of thermal radiation reaches its peak varies inversely with the temperature. Thus, when a body is heated it first glows in the invisible infrared. As the temperature increases, it next becomes red hot, then yellow, white, and finally blue–white, as the peak shifts through the spectrum and on out into the invisible ultraviolet. By the end of the century, physicists Otto Lummer (1860–1925) and Ernst Pringsheim, Sr. (1859–1917) had established experimentally the shape of the so-called "blackbody" radiation spectrum.

At once, astronomers had gained a way to determine the effective ("surface") temperatures of the Sun and stars! Applying these findings to the Sun, astronomers found the effective temperature to be about 6000 K (Cox 2000; see Figure 2.2). This was a huge improvement over earlier estimates, which had ranged from about 1700 K to more than 30,000 K, and it also made possible the determination of bolometric corrections and thus bolometric magnitudes and luminosities for the stars. Increasingly accurate determinations of the effective temperatures of stars followed during the

---

[7] Called a "blackbody" by physicists because it absorbs all wavelengths of radiation equally. The spectrum of radiation emitted by a blackbody depends only upon its temperature.

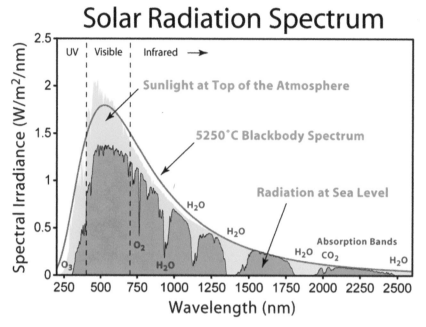

Figure 2.2. The observed solar spectrum, compared with a theoretical calculation of the spectrum of an ideal blackbody at a temperature of 5250 K. The shortest wavelengths, to the left in the figure, correspond to invisible ultraviolet (UV) radiation, the visual band corresponds to visible wavelengths near the peak of the solar spectrum, and increasingly long infrared wavelengths extend to the right. The wavelengths are given in nanometers, with one nanometer equaling 10 Å. Source: commons.wikimedia.org/wiki/File:Solar_Spectrum. png. Attribution: Robert A. Rohde, CC BY-SA 3.0.

20th century, as astronomers learned how to calculate increasingly accurate models of stellar atmospheres, the sources of the observed stellar spectra.

For example, the apparent visual magnitude of Sirius—the "Dog Star," α Canis Majoris—is −1.46, and using the measured parallax of 0.380 seconds of arc (or, 0.380″), which yields a distance of 2.631 pc, astronomers found the absolute visual magnitude to be +1.45. That is, at the standard distance of 10 pc, Sirius would appear to be about 3 magnitudes fainter than it is at its actual, closer, location. With a bolometric correction of about 0.25, corresponding to its effective temperature of about 10,000 K, the bolometric magnitude is +1.21. Note that the absolute bolometric magnitude of a star is *always* smaller than the absolute visual magnitude—meaning that the bolometric measurement is brighter because it provides a measure of *all* of the energy flux from the star, not just the flux in the visual wavelength band. The bolometric magnitude of Sirius is equivalent to a luminosity about 25 times larger than the solar value—or in the logarithmic form that theorists prefer, $\log (L/L_{\mathrm{Sun}}) = +1.4$. The table below lists the values of these same quantities for Sirius, the Sun (http://en.wikipedia.org/wiki/Bolometric_correction), and the red supergiant Betelgeuse (Allen 1955b)—α Orionis, the brightest star in the constellation Orion:

Star	m_V	d (pc)	M_V	M_{bol}	log (L/L_{Sun})	L/L_{Sun}
Sirius	−1.46	2.631	+1.45	+1.2	+1.4	25
Sun	−26.75	4.85×10^{-6}	+4.82	+4.75	0.0	1
Betelgeuse	+0.4	180.	−5.9	−8.0	+ 5.1	130,000

Measurements of the effective temperatures and luminosities of the stars also enabled astronomers to determine stellar radii. From the Stefan–Boltzmann law, the power output per unit area of a star's surface[8] is equal to a known physical constant multiplied by the fourth power of the star's effective temperature. And the product of that quantity by the surface area of the star yields the stellar luminosity—the total power output of the star. Conversely, if we know the luminosity and effective temperature, we can determine the stellar radius. Applying this approach to Sirius, we get a radius a bit more than 1.5 times the solar radius, while for Betelgeuse we find about 1300 times the solar value.

In the 20th century, it became possible to check these radius determinations by direct measurements. In 1920, the Nobel-Prize-winning physicist Albert Michelson (1852–1931) measured the angular diameter of Betelgeuse using a device called an interferometer that he built and installed at the 100-inch telescope on Mount Wilson. He found the angular diameter to be 0.047 seconds of arc (Michelson & Pease 1921), corresponding to a radius about 900 times the solar radius. Decades later, Robert Hanbury Brown (1916–2002) employed the "intensity interferometer" he had invented to measure the radii of some 30 stars, including Sirius, for which he found the angular diameter to be about 5.6×10^{-3} arcseconds (Hanbury Brown & Twiss 1956), corresponding to a radius about 1.6 times the solar value. These values—together with results obtained for other stars—provided direct confirmation that astronomers' measurements of stellar properties were truly yielding physically correct results. Interferometric measurements of stellar diameters and binary separations continue to be made, with steadily increasing accuracy, to the present day.

Such measurements of the observable properties of the stars made possible a direct, graphical comparison between the Sun and the stars during the first decade of the 20th century. Realizing that all stars in clusters such as the Pleiades lie at about the same distance from the Earth, Danish amateur astronomer Ejnar Hertzsprung (1873–1967) recognized that the relative differences in their apparent magnitudes reflected the differences in their absolute magnitudes. When he plotted the absolute magnitudes of the cluster stars against crude measures of their colors—which in turn depend upon their surface temperatures—he found that most stars fell along a diagonal band that he termed the "Main Sequence," a name we use to this day.

---

[8] The observable surface of a star is termed its "photosphere," since it is the source of the observed stellar radiation.

About this same time, Henry Norris Russell (1877–1957), a young faculty member at Princeton University, was independently working on the problem of stellar luminosities using parallax measurements he and others had made to determine the distances to individual stars. He had become friends with Edward C. Pickering (1846–1919), the director of the Harvard College Observatory in Cambridge, Massachusetts, and Pickering provided him with determinations of the spectral classes for the stars for which Russell had obtained distances. During the final decade of the 19th century, Pickering had assigned to Mrs. Williamina P. Fleming (1857–1911), one of the young women on the observatory staff, the task of searching through and classifying hundreds of thousands of photographic images of stellar spectra. With no formal training in astronomy, she simply grouped bright, blue stars like Sirius together in a class she labeled "A." Another group of spectra that were similar to each other she labeled "B," and so on up through the alphabet. Stars like the Sun she termed spectral class G, and she grouped together as spectral class M a group of stars with reddish spectra and strong, dark bands. After the element helium was discovered in the solar spectrum in 1868, the further spectral class O—typified by the presence of lines due to singly ionized helium—was added; the hot, bright blue stars in the constellation Orion belong to this class. In 1896, Pickering added Annie Jump Cannon (1863–1941) to the staff at the Harvard College Observatory. With some graduate-level training in astronomy, she was able to simplify Mrs. Fleming's classification system, retaining only the spectral classes O, B, A, F, G, K, and M—known to generations of astronomy undergraduates through the mnemonic "Oh Be A Fine Girl/Guy, Kiss Me"—while adding decimal subdivisions to accommodate intermediate spectral types. In Cannon's revised system, for example, the Sun is classified as belonging to spectral class G2.

In 1914, by plotting the stellar luminosities derived from the distance measurements he had acquired against the Harvard spectral classes, Russell found essentially the same type of diagonal "Main Sequence" band that Hertzsprung had first discovered, as well as a number of stars that lay above the Main Sequence, at redder colors (and thus lower surface temperatures) and higher luminosities (Figure 2.3). Since stars with higher luminosities at the same or lower temperatures must have larger radii (from the Stefan–Boltzmann law), Russell called them "giants," while the Main Sequence stars he termed "dwarfs." The few white dwarf stars (see Chapter 15) known by the middle of the second decade of the 20th century occupied a lonely position by themselves at lower luminosities than any other stars with similar colors or spectra. This "Hertzsprung–Russell diagram" has since become an important tool for astronomers, providing a map of the relationships between different phases of stellar evolution. Various forms of the "H–R" diagram that have been utilized for different purposes by observers and theoreticians are summarized in Appendix A.

Because the Sun was thus established as a perfectly ordinary Main Sequence star, it made sense for astronomers to adopt the physical properties of the Sun as a convenient "yardstick" to use in measuring the properties of other stars. For convenient reference, the current values of some of the solar quantities are listed in Appendix B.

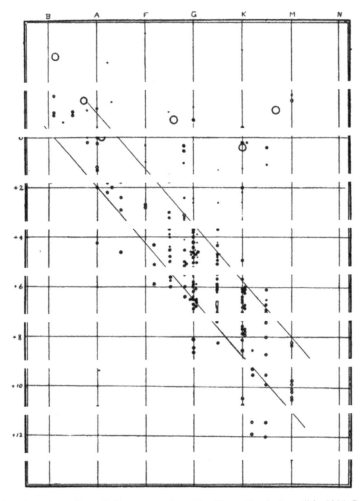

Figure 2.3. The Hertzsprung–Russell diagram as plotted by Henry Norris Russell in 1914. Reprinted from Nature, **93**, 252, copyright 1914; image used under license number 5490360074871 from Springer Nature. Points represent individual stars, which are plotted by spectral type (ranked left-to-right from hottest to coolest) and by absolute magnitude (from brightest at the top to faintest at the bottom). The Main Sequence is represented by the stars between the two diagonal lines, while red giants populate the upper right-hand part of the diagram. The lone white dwarf in this plot, 40 Eridani B, is the isolated point in the lower left part of the figure. The Sun would occupy a position about in the middle of the diagram, with absolute magnitude about +5 on the vertical row immediately to the left of the line labeled "G."

References

Aitken, R. G. 1918, The Binary Stars (New York: D. C. McMurtrie)

Allen, C. W. 1955a, Astrophysical Quantities (2nd ed.; Univ. London: Athlone Press) 225

Allen, C. W. 1955b, Astrophysical Quantities (2nd ed.; Univ. London: Athlone Press) 229

Cox A. N. (ed) 2000, Allen's Astrophysical Quantities (New York: AIP/Springer) 341

Hanbury Brown, R., & Twiss, R. Q. 1956, Natur, 178, 1046; see also

    Hanbury Brown, R., Davis, J., & Allen, L. R. 1974, MNRAS, 167, 121

Hertzsprung, E. 1923, BAN, 2, 15

http://en.wikipedia.org/wiki/Bolometric_correction; accessed 26 November 2014

MacRobert, A. 2006, S&T, www.skyandtelescope.com/astronmy-resources/the-stellar-magni-tude-system

Michelson, A. A., & Pease, F. G. 1921, ApJ, 53, 249

Pannekoek, A. 1961a, A History of Astronomy (London: George Allen & Unwin Ltd.) 342

Pannekoek, A. 1961b, A History of Astronomy (London: George Allen & Unwin Ltd.) 430

Chapter 3

The Role of Radiation Inside Stars

The mathematical model developed by J. Homer Lane in 1870, as well as other 19th-century polytropic models, indicated that the internal temperatures of the Sun and stars reach several million degrees Kelvin. While these models employed the best physics of the day, they did not include the effects of the thermal radiation ("blackbody radiation") in the stellar interiors. It took another half century of developments in the physical sciences before our knowledge of the stars could advance further.

Understanding the properties of blackbody radiation had become an active area of research in the final decades of the 19th century. At the turn of the century, German physicist Max Planck (1858–1947), in seeking to calculate the radiation spectrum theoretically, made the ad hoc assumption that the radiation is emitted in the form of discrete packets, or "quanta," today called "photons." After the experimental verification of Planck's quantum theory of radiation a few years later, astronomers were able to calculate the energy density of thermal radiation and the pressure that it exerts. Both are proportional to the fourth power of the absolute temperature.

Sir Arthur Stanley Eddington (1882–1944; see Figure 3.1) was the first person to incorporate radiation into the calculation of stellar models. Born in England and raised a Quaker, Eddington had become chief assistant to the Astronomer Royal in 1906 at the young age of 24, and his outstanding abilities led to his promotion to Plumian Professor of Astronomy and Experimental Philosophy seven years later (http://en.wikipedia.org/wiki/Arthur_Eddington). Eddington had become fascinated with the problem of understanding the internal structures of the stars, and in 1916, he demonstrated the importance of the pressure exerted by thermal radiation in regulating their internal properties. As he put it in his classic 1926 treatise, *The Internal Constitution of the Stars*, familiarly known as *ICS* (Eddington 1926a), "It is necessary to take account of a phenomenon ignored in the early investigations, viz. the pressure of radiation…. Ordinarily the pressure of radiation is extremely minute

3-1

Figure 3.1. Sir Arthur Stanley Eddington. Source: AIP Emilio Segré Visual Archives, Gift of Subrahmanyan Chandrasekhar; copyright American Institute of Physics.

and can only be demonstrated by very delicate terrestrial experiments; but the radiation inside a star is so intense that the pressure is by no means negligible as regards the conditions of equilibrium of the material" (Eddington 1926b). He went on to give the following very picturesque illustration of the effects of radiation pressure: "We can imagine a physicist on a cloud-bound planet who has never heard of stars calculating the ratio of radiation pressure to gas pressure for a series of globes of gas of various sizes..." Eddington went on to tabulate the fractions of the total pressure provided by both radiation pressure and gas pressure for spheres of gas of various masses in equilibrium, and he found that for masses less than about 10^{32} grams (less than a tenth the mass of the Sun), radiation pressure accounts for less than 1% of the total, while for objects more massive than about 10^{36} grams (more than 500 times the solar mass), it accounts for more than 95%. Between these two limits, he pointed out, "we may expect something interesting to happen. What happens is the stars."

Eddington accordingly modified the differential equations that scientists use to calculate the hydrostatic equilibrium of a star to incorporate the effects of radiation pressure. Specifically, he assumed that the ratio of the radiation pressure to the total pressure (equal to the sum of the radiation pressure plus the gas pressure) is constant throughout a particular stellar model, although it could take on different values in different models. With this assumption, Eddington was able to write the total pressure as a constant multiplied by the density raised to the 4/3 power; that is, the model is a polytrope of index $n = 3$. For a polytrope with this index, the mass is a unique constant, which depends upon the mean molecular weight[1] of the gas and the ratio of the radiation pressure to the total pressure. For a given value of the mean molecular weight, a specific value of this ratio thus yields a specific value of the stellar mass. Eddington found (Eddington 1926c) that if the radiation pressure is only a fraction 0.001 of the total pressure the stellar mass is 0.1284 M_{Sun}, while a value of 0.05 produces 1.004 M_{Sun}, and 0.80 yields 90.63 M_{Sun}.

Eddington's calculations also provided explicit values for the distribution of temperature inside the Sun (see Figure 3.2). The blue curve in this figure shows the actual values from these calculations. For comparison, the yellow curve shows the internal temperature distribution in a model for the Sun computed by the late Princeton astrophysicist John N. Bahcall (1934–2005) and his collaborators in 2005 (Bahcall, et al. 2005). The most obvious difference is that the central temperature in Eddington's model is almost triple the value of one of the very best recent models. This turns out to be in part an artifact resulting from the rudimentary and very poor knowledge of the chemical composition of the Sun at the time of Eddington's work. In brief, because of the prevailing bias among astronomers at the time that hydrogen was unlikely to constitute a significant fraction of the Sun's mass—which we now know to be false—Eddington adopted for the mean molecular weight the value 2.1, in units of the mass of the hydrogen atom. In contrast, for Bahcall's solar model the value instead is 0.590. It turns out that the central temperature of the Sun is directly proportional to the assumed value of the mean molecular weight, and scaling Eddington's calculation by the ratio of Bahcall's value to the one Eddington had assumed gives the dotted pink curve in Figure 3.2; it agrees very much better with modern values.

---

[1] The mean molecular weight is the average mass of the independent particles in a gas. This includes the very-low-mass electrons, which are liberated in abundance in the highly ionized interiors of stars. For example, in units of the mass of a hydrogen atom, the mean molecular weight of a fully ionized, pure H gas is ½ (one unit of mass per two particles, a proton and an electron, neglecting the mass of the electron); for fully ionized He it is 4/3 (four units of mass—the mass of a helium atom—per three particles, two electrons and a helium nucleus); while for all heavier elements it is approximately 2.

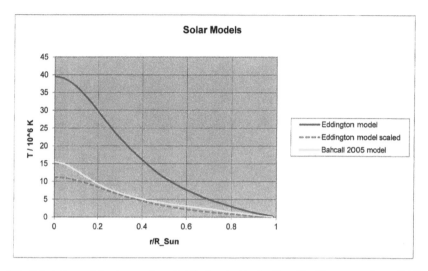

Figure 3.2. Calculations of the temperature distribution inside the Sun. The blue curve shows Eddington's "Standard Model," a polytrope with index $n = 3$; the dashed pink curve shows the same model with the temperatures scaled by the ratio of the assumed mean molecular weights (see text); and the yellow curve is Bahcall's (2005) "Standard Solar Model."

However, not only does radiation provide pressure to help support a star against its own self-gravity, but also it competes with convection to transport heat outward from the place where it is produced deep in the stellar interior to the stellar surface, where it is radiated away into space as starlight.

In addition to including the effects of radiation pressure upon the hydrostatic equilibrium of a stellar model, Eddington also made the first effort to take into account the transport of energy associated with the flow of radiation outward from the center of a star. That is, as Eddington explains, he replaced "Lane's hypothesis of *convective* equilibrium by *radiative* equilibrium. In all the earlier researches it was supposed that the heat was carried from the interior to the surface of the star by convection currents, so that the interior was kept thoroughly stirred and followed the same law of thermal equilibrium as the lower part of the Earth's atmosphere. But it appears now that the heat is transferred by radiation and the temperature distribution is controlled by the flow of radiation; convection currents, if they exist, will strive to establish a different distribution, but the temperature continually slips back to radiative equilibrium since the transfer by radiation is much more rapid" (Eddington 1926d). Eddington accordingly made use of an equation that governs the transport of radiation. In this equation, the gradient of the radiation pressure is proportional to the radiation flux multiplied by a quantity called the "opacity." The opacity regulates the rate of energy flow, as a medium with very low opacity is effectively transparent to radiation, while in one with high opacity, radiation diffuses outward very slowly.

To proceed further, astrophysicists had to be able to calculate the opacity of the matter inside a star and the rate at which energy is generated in the deep interior. Both were made possible by continuing advances in scientists' understanding of the nature of matter and of the interaction between matter and radiation that were being discovered by physicists in the early decades of the 20th century. In 1913, Danish physicist Niels Bohr (1885–1962) developed the first quantum model for the atom. Taking the electrons to be bound in orbits around the positively charged nucleus by the electrostatic forces between them—similar to the way planets are bound into orbits about the Sun by the force of gravity—Bohr assumed that only certain discrete ("quantized") orbits were allowed. He argued that when an atom absorbs a quantum of light (a "photon") having just the right amount of energy, the electron jumps from an allowed lower-energy orbit to an allowed higher-energy orbit. Because the photon energy is directly proportional to its frequency, the absorption of light by the atom produces a decrease in radiation at that frequency, creating a dark absorption line in the spectrum. Conversely, if an electron in a higher-energy state drops down to a lower-energy orbit, it emits a photon with a frequency proportional to the difference in energy between the two states. Applying these ideas to the hydrogen atom—the simplest of all elements, consisting of a single electron orbiting a single, positively charged proton—Bohr found that transitions between the quantized levels produced exactly the series of hydrogen absorption or emission lines first identified in 1885 by physicist Johann Jakob Balmer (1825–1898).

Bohr also found that there is one electron orbit that has the lowest energy for that atom; it is called the "ground state." Since electrons in higher-energy orbits spontaneously emit photons, either in one "quantum jump" or in a sequence of jumps, an isolated atom—removed from any exposure to radiation—ultimately drops into the ground state and stays there. In addition, Bohr found that there is a maximum energy that an electron can have and still occupy a bound orbit in a given atom. The difference in energy between the ground state and this level with maximum energy is called the "ionization potential" of the atom. If the atom absorbs a photon with a greater energy than this, the negatively charged electron is ejected from the atom, leaving behind a positively charged nucleus. The process of removing electrons from an atom is called "ionization," and the positively charged nucleus left behind is termed an "ion." For atoms of chemical elements other than hydrogen—and for heavier isotopes of hydrogen itself—the nucleus consists of some combination of positively charged protons and electrically neutral particles called "neutrons." For example, the nucleus of the helium atom consists of two protons and two neutrons tightly bound together by a force called the "strong nuclear force."

However, if an atom is immersed in a "bath" of radiation, from which it can absorb photons or into which it can emit them, some of the higher-energy states will become populated as well. Electrons continuously jump into and out of these states, but in equilibrium—where on average as many are jumping into a higher state as are jumping out of it—a calculable fraction of the higher states is statistically occupied. As Eddington picturesquely put it (Eddington 1926e), "The inside of a star is a hurly-burly of atoms, electrons and aether waves [N.B., photons]. We have to call to aid the most recent discoveries of atomic physics to follow the intricacies of the dance. We started to explore the inside of a star; we soon find ourselves exploring the inside of an atom. Try to picture the tumult! Dishevelled atoms tear along at 50 miles a second with only a few tatters left of their elaborate cloaks of electrons torn from them in the scrimmage. The lost electrons are speeding a hundred times faster to find new resting-places. Look out! There is nearly a collision as an electron approaches an atomic nucleus; but putting on speed it sweeps round it in a sharp curve. A thousand narrow shaves happen to the electron in 10^{-19} of a second; sometimes there is a side-slip at the curve, but the electron still goes on with increased or decreased energy. Then comes a worse slip than usual; the electron is fairly caught and attached to the atom, and its career of freedom is at an end. But only for an instant. Barely has the atom arranged the new scalp on its girdle when a quantum of aether waves run into it. With a great explosion the electron is off again for further adventures. Elsewhere two of the atoms are meeting full tilt and rebounding, with further disaster to their scanty remains of vesture."

Look again at Figure 3.2; as one proceeds inward from the surface of the Sun, the local temperature increases rapidly, and the energies of thermal photons increase as well. At the center of the Sun, temperatures of several million degrees Kelvin prevail. Accordingly, inside the Sun and the stars "the radiation consists of X-rays .... For example, at 10,000,000° the radiation is mainly between 3 Å and 9 Å wave-length, and it is the absorption coefficient of matter for radiation of this kind that we have to study" (Eddington 1926f).

There are four main contributions to the opacity in the interior of a star: (1) "bound-bound"—line—absorption in which an atom in a lower-energy state absorbs a photon and jumps to a higher-energy state; (2) "bound-free" absorption— also called "photoionization"—that results from the absorption of a photon by a bound electron and its ejection from the atom or ion (the "great explosion" in the above quotation from Eddington); (3) "free-free absorption" produced by the absorption of radiation by an electron passing by a positively charged nucleus (the "side-slip at the curve," as an electron speeds around an atomic nucleus); and (4) scattering by the free electrons that are liberated by ionization.

In 1923, the Dutch physicist Hendrik A. Kramers (1894–1952) developed a physical quantum theory for the absorption of radiation by the bound electrons in atoms and ions (Kramers 1923). Kramers' theory agrees very well with laboratory experiments and has a form that can immediately be applied to matter in the deep interiors of stars. The following year, the Norwegian theoretical astrophysicist Svein Rosseland (1894–1985) showed (Rosseland 1924) that the most directly useful form of the opacity in the deep interior of a star is the harmonic mean, averaged over the gradient of the thermal radiation spectrum (as represented by Planck's expression for the spectrum of blackbody radiation). This is still termed the "Rosseland mean opacity." The reason for this form of the opacity is simple: Radiation flows outward most easily through regions of the spectrum in which the opacity is low—which receive higher weight when divided by the reciprocal of the frequency-dependent opacity—while radiation flow is greatly reduced for regions of the spectrum in which the opacity is high—and which receive correspondingly lower weight from the reciprocal opacity.

Eddington applied these then-new quantum calculations of the opacity of stellar matter in the work described in his landmark book. Because the source of stellar energy was unknown at the time of Eddington's work, however, he approximated the radiant energy flux by defining an (unknown) dimensionless function that expressed the energy per second flowing outward through a spherical surface that contains a specific amount of mass, divided by the luminosity at the stellar surface. The resulting equation is proportional to the product of this quantity and the opacity, and to solve the equations, Eddington assumed this product to be a constant throughout the star. This became known as his "standard model" for a star (Chandrasekhar 1939). Solving the standard model enabled Eddington to express the radius, central temperature and density, and luminosity of a stellar model in terms of the stellar mass, effective temperature, mean molecular weight, and the assumed ratio of radiation pressure to total pressure. If the mean molecular weight is a fixed constant, the mass becomes a unique function of this ratio, and it is then possible to combine the results of the calculations to find a unique relation between the mass and luminosity of a star. Eddington fixed the constant of proportionality by requiring that the model calculations yield the known value of the luminosity of the nearby star Capella at a mass equal to its known mass. He was then able to calculate explicit values for the mass–luminosity relation, which he plotted as a graph (see Figure 3.3). When he subsequently added data points representing the masses and luminosities for individual stars (also shown in Figure 3.3), he found amazingly good agreement with his theoretical results.

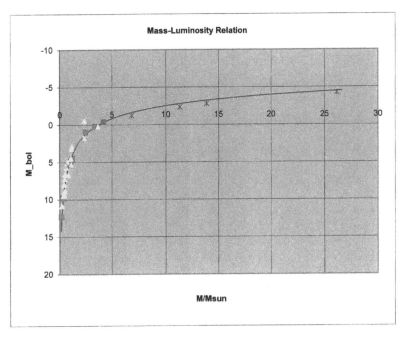

Figure 3.3. Eddington's mass–luminosity relation for Main Sequence stars. All data are from Eddington (1926); theory, p. 137, and observations, p. 154. The solid curve represents Eddington's theory, while the individual symbols represent the observations; the pink squares and yellow triangles represent ordinary binary stars, the blue crosses are for eclipsing binaries, and the purple asterisks are for Cepheid variables (pulsating stars; see Chapter 20).

As Eddington wrote (Eddington 1926g),

> "The theoretical mass–luminosity law for a perfectly gaseous star was obtained in February 1924 and the comparison with observation... followed. The agreement of the observations with the curve was a complete surprise for it was not at all the result that was looked for."

This quotation clearly captures the excitement and gratification any theorist feels when his or her predictions are tested against observation or experiment and agree!

Despite the substantial progress Eddington had made in understanding the nature of the stars, two major problems remained by the end of the second decade of the 20th century: (1) What are the chemical compositions of the Sun and the stars? (2) And what constitutes the source of their tremendous energy? We shall discuss the answers to these questions in the next two chapters.

References

Bahcall, J. N., Serenelli, A. M., & Basu, S. 2005, ApJ, 621, L85

Chandrasekhar, S. 1939, An Introduction to the Study of Stellar Structure (Chicago, IL: Univ. Chicago Press) 228

Eddington, A. S. 1926a, The Internal Constitution of the Stars (Cambridge: Cambridge Univ. Press) 8

Eddington, A. S. 1926b, The Internal Constitution of the Stars (Cambridge: Cambridge Univ. Press) 15

Eddington, A. S. 1926c, The Internal Constitution of the Stars (Cambridge: Cambridge Univ. Press) 137

Eddington, A. S. 1926d, The Internal Constitution of the Stars (Cambridge: Cambridge Univ. Press) 9

Eddington, A. S. 1926e, The Internal Constitution of the Stars (Cambridge: Cambridge Univ. Press) 19

Eddington, A. S. 1926f, The Internal Constitution of the Stars (Cambridge: Cambridge Univ. Press) 21

Eddington, A. S. 1926g, The Internal Constitution of the Stars (Cambridge: Cambridge Univ. Press) 164

http://en.wikipedia.org/wiki/Arthur_Eddington; accessed 3/2/15

Kramers, H. A. 1923, PMag, 46, 836

Rosseland, S. 1924, MNRAS, 84, 525

Chapter 4

The Nature of Star Stuff

Today, we take it as given that the Sun and stars are giant balls of hydrogen and helium gas, with a smattering of heavier elements thrown in. It is thus something of a shock to realize that this understanding came about only around the middle of the 20th century.

The 1814 discovery by Joseph von Fraunhofer (1787–1826) that the solar spectrum contains dark lines at many different wavelengths provided the first clue that ultimately enabled astronomers to determine the chemical compositions of the Sun and stars. Although Fraunhofer had no idea what caused the lines, physicists and astronomers worked hard over the next half century to find out. By the time Gustav Kirchoff (1824–1887) formulated his three laws of radiation in the 1860s, scientists had learned that groups of these dark lines were the "fingerprints" of individual chemical elements. Armed with this knowledge, they soon determined that the gas at the visible surface of the Sun consists primarily of neutral atoms, together with some ions from a few elements—such as calcium (Ca), sodium (Na), or iron (Fe)—that are readily ionized by thermal photons at the 5800 K effective temperature of the Sun.

However, "In spite of the widespread awareness that the solar surface was hot enough to vaporize any known terrestrial material, by 1870 there was still near unanimity that the photosphere [the visible surface of the Sun] contained clouds full of liquid or solid particles... The suggestion that these clouds consisted of carbon particles is commonly attributed to G. Johnstone Stoney, from a paper that appeared in 1867.... [And], writing in 1872, Richard Pryor insisted that [matter] 'probably ... [exists] in certain portions of the solar globe as liquids or even as solids'" (Powell 1988).

Interest in the chemical composition of heavenly bodies was not confined just to the Sun. In 1863, "William Huggins in his private observatory at Tulse Hill in England ... studied the spectra of a number of bright stars, identified the dark lines of sodium, iron, calcium, and magnesium, and stated … that the same elements are present in the stars as in the Sun and on the Earth" (Pannekoek 1961a).

Demonstrating from their spectra that the Sun and stars consist of the very same chemical elements that occur on Earth was an important step forward in understanding the nature of these celestial objects, but it was not sufficient to enable astrophysicists to construct physical models. Not only did they need to know *what* the Sun and stars are made of, but also they needed to know *how much* of each chemical element was present. Even though astronomers could see the absorption lines produced by neutral hydrogen in the spectra of the Sun and stars—and consequently knew that some amount of hydrogen must be present—the idea that the Sun might consist primarily of heavy elements proved extremely hard to shake.

Even half a century after J. Homer Lane's pioneering work began the development of a physical model for the Sun, British astrophysicist Arthur Eddington stated (Eddington 1926a), "… we have in mind as the most likely constitution of the material [in stars] a predominance of elements in the neighborhood of iron with some admixture of lighter non-metallic elements…" Speculating that the source of solar energy seemed likely be the fusion of hydrogen nuclei to form helium—by some then-unknown process, which evidently necessitated *some* abundance of hydrogen in the solar interior—Eddington continued (Eddington 1926b), "An initial proportion of 7% of hydrogen is necessary for a life of 10^{10} years… [but] We should be reluctant to admit a greater proportion…"

At almost exactly the same time that Eddington was penning these words, back on the other side of the Atlantic, a brilliant young woman, Cecilia H. Payne (1900–1979; Figure 4.1), was uncovering evidence that the prevailing ideas about the internal compositions of the stars were completely wrong and that they are in fact composed primarily of hydrogen and helium.[1] Born in England during the first year of the 20th century, Cecilia Payne was one of three siblings. The year after the end of the Great War (renamed World War I after the second global conflict began in 1939), she entered Cambridge University to study science (http://en.wikipedia.org/wiki/Cecilia_Payne-Gaposchkin). Early in her college studies, she attended a lecture by Eddington, well-known as a compelling speaker, which ignited in her a passion for astronomy and astrophysics that burned brightly for the rest of her life. Because Cambridge did not grant degrees to women until 1948, Miss Payne realized that she would not be able to pursue a scientific career in England. When Harlow Shapley (1885–1972), then the Director of the Harvard College Observatory, offered her a fellowship to carry out graduate studies in America, she jumped at the chance, leaving England in 1923.

Shapley, who with Eddington served as Cecilia's Ph.D. thesis advisor, encouraged her to work on the analysis of stellar spectra for her doctoral research. This made good use of the treasure trove of such data that had been amassed at the Observatory. Shapley's predecessor, Edward C. Pickering (1846–1919), had employed a staff of young women to go through the thousands of stellar spectra that had been painstakingly acquired and group them into classes with similar

---

[1] The first draft of this chapter was completed before Donovan Moore's excellent biography of Cecelia Payne-Gaposchkin: Moore (2020).

Figure 4.1. Cecilia H. Payne. Source: Smithsonian Institution from United States, no restrictions, via Wikimedia Commons.

characteristics. Annie Jump Cannon (1863–1941), one of the most accomplished of these assistants, established the system of spectral classification that astronomers still use today. She classified the hottest, blue stars with absorption lines produced by singly ionized helium as spectral class O. Slightly cooler were the B stars, which display lines of neutral helium. The white-hot A stars, like Sirius, show strong absorption lines due to hydrogen. Then came the cooler F and yellow G stars—like the Sun—which exhibit lines of metallic elements. The orange K stars and red M stars, which contain absorption features produced by molecules like CO, H_2O, HCN, SiO, etc., are the coolest of these spectral classes.

At the time Cecilia Payne began her doctoral research, physicists and astrophysicists around the globe were just beginning to develop an understanding of the atmospheres of the Sun and stars based on applications of quantum theory and an understanding of the transport of radiation through ionized gases. "Ionization theory became fertile for astrophysics when in 1920 the Bengali physicist Megh Nad Saha (1893–1956) derived the ionization formula, expressing the … ionization of any atom as a result of temperature and pressure. Now the intensity of a spectral line became a quantity calculable and dependent on the physical conditions of the stellar atmosphere that produced it; conversely, these physical conditions in a star can be derived from the line intensities in its spectrum. Thus the ionization formula became the basis of a quantitative treatment of stellar spectra. Saha could at once explain the differences in the consecutive spectral classes indicated as O, B, … K, M, by differences in temperature…. Now it was evident that chemically equal atmospheres, if increasing in temperature, had to show the sequence M, K, G, F, A, B, O spectra" (Pannekoek 1961b).

Payne's Ph.D. thesis (Payne 1925a), published in 1925, summarized the existing physical basis for the quantitative analysis of stellar spectra, focusing on Saha's theory of thermal ionization of the elements and on its applications to the theory of stellar atmospheres then being developed by Fowler, Milne, Eddington, and others. She went on to derive a large amount of observational data from the thousands of Harvard spectra, including estimates of the intensities of absorption lines across the different spectral classes, and she derived her own scale of effective temperatures for the different spectral classes.[2] With this background, she derived a number of physical quantities for the stars from their spectra, addressed several cases with notable spectral peculiarities, and ended by deriving quantitative estimates for the relative abundances of the chemical elements. It is no wonder that astronomers Otto Struve and Velta Zebergs referred to her thesis as "undoubtedly the most brilliant Ph.D. thesis ever written in astronomy" (http://en.wikipedia.org/wiki/Cecilia_Payne-Gaposchkin).

Two years before Payne completed her doctoral work, astrophysicists Ralph H. Fowler (1889–1944) and Edward Arthur Milne (1896–1950) had pointed out that the strength of an absorption line in a stellar atmosphere depends upon the abundance of the chemical element producing the line as well as upon the effective temperature and surface gravity (Fowler & Milne 1923). Payne made use of their approach to derive the abundances of the chemical elements in stars from the appearances of various absorption lines in the stellar spectra. Acknowledging that her approach was likely to provide only a first approximation to the actual abundances, and recognizing the prevailing opinion that stars were then still thought to be composed largely of heavier elements, she wrote (Payne 1925b), "The stellar abundance for … [hydrogen and helium] is improbably high, and is almost certainly

---

[2] In comparison with current values, Payne underestimated the temperatures of stars cooler than the Sun (T_{eff} = 5800 K) by up to about 1,000 K; overestimated those up to the cooler B stars (T_{eff} between about 6000 K up to about 12,000 K) by a similar amount or more; and again underestimated the temperatures of the hottest stars, with the differences growing to as much as 10,000 K for the O stars (T_{eff} greater than about 34,000 K). Given the substantial improvements in both observation and theory since this early work, this level of agreement is remarkably good, and it provides a strong testimonial to Payne's abilities.

not real." Four years later, while maintaining his reservations about her determinations of the abundances of these two light elements, Henry Norris Russell (Russell 1929) nevertheless wrote, "The most important previous determination of the abundance of the elements by astrophysical means is that by Miss Payne..." Comparing his latest results with hers, Russell commented further, "This is a very gratifying agreement, especially when it is considered that Miss Payne's results were determined by a different theoretical method, with instruments of a quite different type (Harvard objective prism), and even on different bodies—a long list of stars, almost all of which are giants. About the only common feature are the observations of spectral lines and the use of the ionization theory." Figure 4.2 shows a comparison between the logarithms of the abundances determined by Payne in 1925, by Russell in 1929, and by modern determinations (Cox 2000). The general agreement between the three—spanning three-quarters of a century!—is truly remarkable. It justifiably establishes Payne as having provided the first quantitative estimates of the relative abundances of the elements and thus establishing that the compositions of the stars are dominated by hydrogen and helium rather than the heavier elements. Important differences remained, however. Payne's abundance estimate for hydrogen was almost nine times larger than the current value, while Russell's was more than 40% smaller. For helium, Payne's result was less than 20% of the current value, while Russell did not determine the helium abundance.

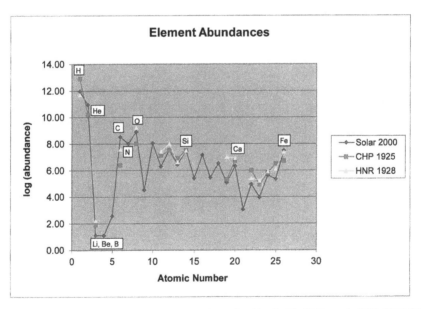

Figure 4.2. Abundances of the chemical elements, as determined by Cecilia H. Payne in 1925 (CHP 1925; the pink data points), Henry Norris Russell in 1929 (HNR 1929; the yellow data points), and solar system values as listed by Cox 2000, p. 29 (the blue data points). The horizontal axis is the atomic number of the element, and the vertical axis is the logarithm of the relative abundance, normalized arbitrarily to 7.50 for silicon (Si). Several of the elements are labeled by their chemical symbols.

Dr. Payne became an American citizen in 1931. Two years later, she met Russian astrophysicist Sergei I. Gaposchkin (1898–1984) and helped him to emigrate to the United States. They were married in 1934, and she became Cecilia Payne-Gaposchkin. She bore him three children, while continuing her productive career in astronomy and collaborating with her husband on observations and analyses of variable stars. After serving for more than a decade as Shapley's assistant, she was granted the title "Astronomer" in 1938. The significance of her work was recognized in 1943, when she was elected to the American Academy of Arts and Sciences. In 1956, she was promoted to full professor in the Faculty of Arts and Sciences at Harvard, and that same year she became Chair of Harvard's Department of Astronomy, the first woman to lead a Harvard department. She continued to conduct active research in astronomy and astrophysics for the remainder of her life. During the course of her career, Payne-Gaposchkin analyzed spectra of high-luminosity stars, reported more than three million variable-star observations (which provided the foundation for their use in studying the structure of the Milky Way Galaxy), and conducted studies of the properties of galactic novae (http://en. wikipedia.org/wiki/Cecilia_Payne-Gaposchkin).

It took quite some time—and considerable additional effort by many astronomers and astrophysicists—before Payne's results indicating that stars are composed primarily of hydrogen and helium came to be accepted. Even as recently as 1939 —after support for Payne's results had begun to be obtained by leading astronomers such as Russell and Strömgren (Strömgren 1932, 1933)—the eminent theoretical astrophysicist Subramanyan Chandrasekhar, in his seminal book on stellar structure, was still extremely cautious about accepting a larger value for the hydrogen abundance. He wrote (Chandrasekhar 1939), "Strömgren has computed the values of … [the mean molecular weight] and has thus inferred the hydrogen content of those stars for which there is reliable information concerning…" their mass, radius, and luminosity. Strömgren's method gave for Capella 0.30 grams of hydrogen per gram of stellar matter, while for the Sun the comparable value was 0.36. This ratio is termed the "mass fraction" of hydrogen and is commonly denoted by the symbol X in astronomy. The comparable mass fraction of helium is denoted by the symbol Y, while the mass fraction of all the heavier elements together is sometimes denoted by Z, which by definition equals $1 - X - Y$. For comparison, the values of the mass fractions of hydrogen, helium, and the remaining heavier elements calculated from the solar system abundances tabulated by Cox (2000) and plotted in Figure 4.3 are $X = 0.71$, $Y = 0.27$, and $Z = 0.02$.

Astrophysicists' conclusions about the internal composition of the Sun continued to vary widely even into the 1950s (see Figure 4.3). The issue was finally resolved only in the following decade. In 1960, Leo Goldberg (1913–1987) and his colleagues published an extensive list of element abundances for the Sun based on spectroscopic observations analyzed with the aid of detailed numerical models for the solar atmosphere (Goldberg et al. 1960). It is not possible to determine the abundance of helium in the Sun spectroscopically, however, because the temperature of the solar photosphere is much too low for thermal photons to be able to produce absorption lines for this element. This problem also was solved in the early 1960s, when

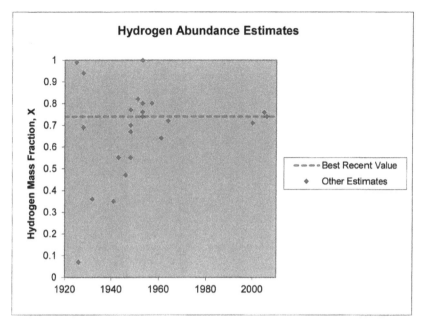

Figure 4.3. Estimates of the hydrogen abundance in the Sun over the past century. The dashed red horizontal line is the best recent determination of the hydrogen mass fraction, $X = 0.7392$, from Asplund, Grevesse, and Sauval 2006. The blue diamonds provide the estimates obtained from various sources during the past century. Not until the middle of the 20th century did the hydrogen abundance determinations begin to converge to a value approaching that accepted today.

spacecraft measurements of the composition of the solar wind determined the helium abundance directly for the surface layers of the Sun. The result was $X = 0.72$, $Y = 0.26$, and $Z = 0.02$ (Gaustad 1964). More recent values are $X = 0.7392$, $Y = 0.2486$, and $Z = 0.0122$ (Asplund et al. 2006), which clearly demonstrate the dominant abundances of hydrogen and helium and the small amounts of all the heavier elements combined.

In other words, hydrogen accounts for almost 74% of the entire mass of the Sun; helium comprises almost 25% more; and all the rest of the heavier elements combined contribute just a little more than 1%. This is a very far cry from the 7% abundance of hydrogen that Eddington and other distinguished scientists believed at the end of the first quarter of the 20th century. Nevertheless, even though only tiny amounts of the heavier elements are present, as we shall see they have very important effects on the properties of matter and the physical processes that occur in the Sun and other Main Sequence stars.

References

Asplund, M., Grevesse, N., & Sauval, A. J. 2006, NuPhA, 777, 1

Chandrasekhar, S. 1939, An Introduction to the Study of Stellar Structure (Chicago, IL: Univ. Chicago Press) 276

Cox A. N. (ed) 2000, Allen's Astrophysical Quantities (New York: AIP/Springer) 29

Eddington, A. S. 1926a, The Internal Constitution of the Stars (Cambridge: Cambridge Univ. Press) 250

Eddington, A. S. 1926b, The Internal Constitution of the Stars (Cambridge: Cambridge Univ. Press) 293

Fowler, R. H., & Milne, E. A. 1923, MNRAS, 81, 403

Gaustad, J. E. 1964, ApJ, 139, 406

Goldberg, L., Müller, E. A., & Aller, L. H. 1960, ApJS, 5, 1

http://en.wikipedia.org/wiki/Cecilia_Payne-Gaposchkin; accessed 1/23/15

Pannekoek, A. 1961a, A History of Astronomy (London: George Allen & Unwin Ltd.) 450

Pannekoek, A. 1961b, A History of Astronomy (London: George Allen & Unwin Ltd.) 460

Payne, C. H. 1925a, PhD thesis, (Radcliffe College)

Payne, C. H. 1925b, PNAS, 11, 192

Powell, C. 1988, JHA, 19, 187

Russell, H. N. 1929, ApJ, 70, 11

Strömgren, B. 1932, ZA, 4, 118

Strömgren, B. 1933, ZA, 7, 222

Chapter 5

Discovering the Source of Solar Energy

By the end of the second decade of the 20th century, the major problem remaining in the early study of stellar structure concerned the source of energy needed to supply the prodigious power output of the stars. Helmholtz's 1854 suggestion that a gradual contraction of the Sun under its own self-gravity might be the source of solar energy proved to be completely inadequate. In order to provide sufficient energy to sustain the Sun's luminosity at its present level, our star would have had to contract by an amount equal to its entire radius in only about 30 million years (Pannekoek 1961), far smaller than the age of the Earth, which was already known from geological evidence to be more than a hundred million years old. Neither internal heat nor energy supplied by hypothetical meteoritic bombardment was any more successful; both resulted in similar ages of a few tens of millions of years. Consequently, in 1861 Lord Kelvin concluded (Chandrasekhar 1939) that "neither the meteoric theory of solar heat nor any other natural theory can account for the solar radiation continuing at anything like the present rate for many hundreds of millions of years."

In 1905, a potential new source of energy was revealed by Einstein's famous equation $E = mc^2$. This implies that a single gram of mass—about 1/30 of an ounce! —is equivalent to about 9×10^{20} ergs of energy, or about 25 million kilowatt-hours, sufficient to supply the energy requirements of several thousand homes for an entire year and more than a billion times larger than the amount that could be released by gravitational contraction. If some way could be found to convert a portion of the mass of the Sun into energy, the mystery of the source of solar energy would be solved. In 1911, Ernest Rutherford (1871–1937) demonstrated that most of the mass of an atom resides in a small, positively charged nucleus. To release the vast amount of energy required to sustain the power output of the Sun, it would clearly be necessary somehow to tap the mass–energy of the atomic nucleus, and in the 1920s, Eddington and others speculated that nuclear energy sources might be the answer.

Eddington was perhaps the first astrophysicist to articulate this point, just a decade and a half after Einstein's breathtaking revelation of the tremendous amount of energy available in matter. As Chandrasekhar put it, looking back in 1988 (Eddington 1926; reissued in 1988 with a foreword by Chandrasekhar), "Among Eddington's predictions, that of the source of stellar energy is perhaps the most spectacular. His address, on 1920 August 24 to the British Association meeting in Cardiff, contains some of the most prescient statements in all of astronomical literature (1920, *Observatory*, **43**, 353–5); and it is worth quoting in full".

> Only the inertia of tradition keeps the contraction hypothesis alive—or rather, not alive, but an unburied corpse. But if we decide to inter the corpse, let us frankly recognize the position in which we are left. A star is drawing on some vast reservoir of energy by means unknown to us. This reservoir can scarcely be other than the subatomic energy which, it is known, exists abundantly in all matter; we sometimes dream that man will one day learn how to release it and use it for his service. The store is well-nigh inexhaustible, if only it could be tapped. There is sufficient in the Sun to maintain its output of heat for 15 billion years...
>
> Aston has ... shown conclusively that the mass of the helium atom is ... less than the sum of the masses of the four hydrogen atoms which enter into it—and in this, at any rate, the chemists agree with him. There is a loss of mass in the synthesis amounting to 1 part in 120, the atomic weight of hydrogen being 1.008 and that of helium just 4.... Now mass cannot be annihilated, and the deficit can only represent the mass of the ... energy set free in the transmutation. We can therefore at once calculate the quantity of energy liberated when helium is made out of hydrogen. If 5 per cent of a star's mass consists initially of hydrogen atoms, which are gradually being combined to form more complex elements, the total heat liberated will more than suffice for our demands, and we need look no further for the source of a star's energy.

However, it is one thing to be able to identify the *source* of stellar energy but something else entirely to calculate the *rate* at which this energy can be liberated. In the 1920s, no one knew how this might be done.

Quantum physics finally enabled scientists to solve the problem of stellar energy generation. In 1924, French physicist Louis de Broglie (1892–1987) theorized that if light waves sometimes behave as particle-like photons, as Max Planck had postulated at the dawn of the 20th century, then perhaps particles like electrons and protons might sometimes behave as waves (De Broglie 1926). If so, he argued that the wavelengths of such matter waves should be equal to Planck's constant divided by the particle momentum—exactly the same relation that applies to photons.

It did not take long for experimental physicists to put this idea to the test. In 1927, American experimentalists Clinton Davisson (1881–1958) and Lester Germer (1896–1971) showed that a beam of electrons bouncing off a grating formed by the regular lattice of atoms in a crystal of pure nickel produce an interference pattern at a detector. Such interference patterns have been a classic test of wavelike properties ever since people first noticed them being produced from the ripples generated by throwing pebbles into a pond. Indeed, just such an interference experiment was employed by Thomas Young (1773–1829) in 1803 to demonstrate the wave nature of light.

In 1926, Austrian physicist Erwin Schrödinger (1887–1961) formulated a wave equation to describe the behavior of matter waves, and in 1928 the 24-year-old Russian physicist George Gamow (1904–1968; see Figure 5.1)—and independently Ronald W. Gurney (1898–1953) and Edward U. Condon (1902–1974)—used it to describe the process called "alpha decay" in nuclear physics (Gamow 1928; Gurney & Condon 1928). In brief, alpha decay is the spontaneous emission of an alpha particle—the bare nucleus of a helium atom—by a heavy nucleus such as radium-226 (^{226}Ra), which results in its transmutation into a nucleus of radon-222 (^{222}Rn), an isotope that is smaller by two units of positive electrical charge and four units of atomic mass, exactly the quantities carried off by the emitted alpha particle. Gamow idealized the initial nucleus as a deep potential-energy well in which the alpha particle remains confined—an approximation to account for the binding caused by the strong nuclear force, which was only beginning to be understood at the time. He assumed the well to have a finite but very small radius, about the size of an

Figure 5.1. Informal portrait of George Gamow. Credit: AIP Emilio Segré Visual Archives.

atomic nucleus.[1] Outside the well—that is, beyond the very short range of the strong nuclear force—Gamow assumed the potential energy to be the pure Coulomb potential produced by the positive electrical charge of the nucleus, as sketched in Figure 5.2. This produces a potential-energy barrier—the gray space in Figure 5.2— that prevents alpha particles from escaping. If alpha particles really do have wavelike characteristics—as suggested by the red curve oscillating around the horizontal black line representing the alpha-particle energy in Figure 5.2—then there is a calculable probability that the alpha-particle wave can tunnel through this barrier, something that would be completely forbidden by classical mechanics. Outside the barrier, the positively charged alpha particle is repelled by the positive

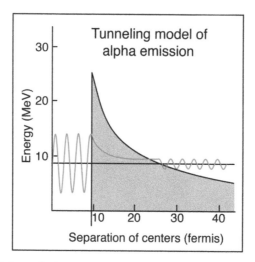

Figure 5.2. Illustration of Gamow's theory of the alpha decay of a radioactive atomic nucleus. The vertical axis gives the energy in millions of electron volts (MeV), while the horizontal axis shows the separation of the alpha particle from the center of the atomic nucleus from which it is being emitted. The horizontal black line at about 9 MeV is the energy of the alpha particle after it has escaped and is far away from the nucleus. The oscillating red curve is intended to suggest the matter wave corresponding to the alpha particle. The gray area outlined by the black curve represents the potential experienced by the alpha particle. The region to the left is the potential well produced by the strong nuclear force that binds the alpha particle to the nucleus, while the gradual decay from the peak represents the electrostatic repulsion between the positively charged nucleus and the positively charged alpha particle. The large amplitude of the oscillating red curve within the nuclear potential well represents the high probability of finding the alpha particle bound within the nucleus. If alpha particles obeyed classical physics—rather than quantum physics—they would not be able to exist at all within the area under the peak of the potential energy, and they would remain trapped within that deep well. The small amplitude of the oscillating red curve to the right of the figure represents the small probability of finding the alpha particle outside the nucleus, i.e., after the original nucleus has decayed by emitting the alpha particle. Source: Reprinted from HyperPhysics with permission from Rod Nave. I am grateful to Rod Nave for supplying a high-resolution version of this figure.

---

[1] The radius of an atomic nucleus is about 10^{-13} centimeters, termed one "fermi" after the brilliant Italian physicist Enrico Fermi (1901–1954), who created the first self-sustaining nuclear fission reaction at the University of Chicago in 1942. This dimension is roughly 100,000 times smaller than the size of an atom.

charge of the residual nucleus and coasts on down the potential hill, emerging at large distances (centimeters, say) as a free particle, represented by the oscillating red curve to the right in the figure. Gamow calculated the quantum-mechanical barrier-penetration probability and found results consistent with experimental measurements on numerous radioactive nuclei.

The process of barrier penetration can also work the other way round, enabling two positively charged nuclei—protons, for example—to tunnel through the repulsive Coulomb potential barrier produced by their like charges and merge together to form a deuteron. Indeed, in 1937 Carl Friedrich von Weizsäcker (1912–2007) suggested this simplest of all nuclear-fusion reactions as a possible source of energy for the Sun:

$$^1H + {}^1H \rightarrow {}^2H + e^+ + \nu.$$

The symbols in this expression represent the particles involved in this nuclear reaction: ^1H is the nucleus of an ordinary hydrogen atom, a proton; and ^2H (sometimes alternatively represented by D) is a deuteron, the nucleus of the "heavy hydrogen" isotope that consists of one proton and one neutron bound together by the strong nuclear force. Because the neutron and proton have approximately the same mass, a deuteron is about twice as heavy as a proton, and this is represented by the superscript "2" in ^2H. The remaining symbols represent a positron (e^+) and a neutrino (ν).[2]

Positively charged particles like the two protons in this reaction tend to repel each other. At the thermal speeds corresponding to the central temperatures of Main Sequence stars, they would therefore never approach each other closely enough to undergo a nuclear reaction. However, Gamow found that this actually can occur because of the quantum-mechanical effect of barrier penetration.

The first calculation of this proton–proton reaction was published in 1938 by Hans Albrecht Bethe (1906–2005; Figure 5.3) and Charles L. Critchfield (1910–1994; Bethe & Critchfield 1938). Because this reaction is mediated by the weak interaction, as indicated by the fact that it involves a neutrino, it is much too slow to be studied in terrestrial laboratories. However, as soon as a deuteron is formed, it rapidly undergoes several further nuclear interactions—mediated by the strong nuclear force—which add more protons. Collectively termed the "proton–proton chain," these reactions ultimately produce the stable nucleus ^4He. They do generate enough energy to power the Sun, but they cannot generate energy rapidly enough to account for the luminosities radiated by the more massive stars.

Hans Bethe was born in Strasbourg, Germany, and he began his university studies in 1924 in Frankfurt. His outstanding abilities—which included not only a brilliant mind but also a willingness and ability to work extremely hard and a truly impressive memory—soon led him to move to Munich. There, he studied under the eminent physicist Arnold Sommerfeld, earning his doctorate in 1928, only four years after starting as a university freshman! He continued his meteoric career as a postdoc at the Cavendish Laboratory in England before becoming a junior faculty member

---

[2] A neutrino is a ghostly particle with no electrical charge and a very small mass, which hardly interacts at all with other forms of matter; see the discussion in Chapter 8.

Figure 5.3. Hans Albrecht Bethe in 1935. Credit: Photograph by Samuel Goudsmit, courtesy AIP Emilio Segré Visual Archives, Goudsmit Collection; copyright American Institute of Physics.

at Tübingen University in 1932. When the Nazis came to power in Germany in 1933, Bethe became one of the casualties of the wave of anti-Semitic academic dismissals. Fortunately, Cornell University in the United States offered him an appointment as an assistant professor, which Bethe accepted, arriving in Ithaca, New York, in 1935.

In 1938, George Gamow organized a conference at the Department of Terrestrial Magnetism of the Carnegie Institution of Washington for a small group of physicists and astrophysicists to discuss the problem of the energy sources of the stars, and Bethe was one of the invited participants. By that time, astrophysicists had already established that the central temperature and density of the Sun are extremely high[3];

---

[3] Eddington's 1926 calculations gave the central temperature and density of the Sun as about 40 million degrees Kelvin and 80 g cm^{-3}, respectively. Following the Carnegie meeting, Bethe adopted as "standard conditions" for the center of the Sun a temperature of 20 million K, density of 80 g cm^{-3}, and hydrogen mass fraction $X = 0.35$ (see Aller, L. H. 1950, ApJ, **111**, 173, "Proton–Proton Reactions in Red Dwarf Stars"). The corresponding modern values from Bahcall's 2005 solar model are 15 million K, 150 g cm^{-3}, and $X = 0.76$.

that the Sun consists mainly of hydrogen and helium; and that the energy-generation rate in more massive Main Sequence stars must be orders of magnitude larger than in the Sun. Bethe was already fascinated with the problem of stellar energy sources, and he thought hard about it during the train ride back from Washington, DC, to Ithaca in light of what he had learned at the conference. According to a popular book by Gamow (Gamow 1945), Bethe did not have the answer by the time the porter came through the train with the first call for dinner. Redoubling his efforts, however, he had the solution before the last call (according to Gamow)! Decades later, after he had received the 1967 Nobel Prize for this work, I asked Bethe about this story. With a grin, he replied, "The truth is that it took me six weeks of hard work in the Cornell library!"

The solution Bethe found (Bethe 1939) was the following sequence of nuclear reactions:

$$^{12}C + {}^1H \rightarrow {}^{13}N + \gamma$$

$$^{13}N + {}^{13}C \rightarrow e^+ + \nu$$

$$^{13}C + {}^1H \rightarrow {}^{14}N + \gamma$$

$$^{14}C + {}^1H \rightarrow {}^{15}O + \gamma$$

$$^{15}O \rightarrow {}^{15}N + e^+ + \nu$$

$$^{15}N + {}^1H \rightarrow {}^{12}C + {}^4He.$$

This sequence of catalytic thermonuclear reactions effectively utilizes the elements carbon (C), nitrogen (N), and oxygen (O) to convert four protons into the nucleus of one helium atom (^4He). In the process, the reactions liberate energy in the form of gamma rays (γ) and positrons (e^+), which are quickly absorbed and converted into local thermal energy, and neutrinos (ν), which escape from the solar interior almost without interacting at all; see Chapter 8. Note that a ^{12}C nucleus starts the process and that a ^{12}C nucleus is regenerated in the last step, ready to start the sequence all over again with another proton.

This sequence of reactions is sometimes called the "Bethe cycle" or the "CNO cycle." Another, interlocking, set of catalytic reactions provides another pathway for combining four protons to produce ^4He and generate energy, and the combination is called the "CNO bi-cycle."

The total energy released by either the p–p chain or the CNO reactions is only about 0.007 of the combined rest-mass energy of the four protons. Although this does not sound like much, it is still sufficient to power the Sun for a very long time. For example, if this amount of energy were released from only 10% of the solar mass, it would be sufficient to sustain the present solar luminosity for about 10 billion years. Radiogenic age measurements have shown that the age of the solar system—including the Sun, the Earth and Moon, and meteorites—is about

4.5 billion years.[4] Thus, the "hydrogen-burning" thermonuclear reactions worked out by Bethe and others are indeed sufficient to power our own star during its entire lifetime.

With calculations of the rates of these and other thermonuclear reactions in hand, astrophysicists thus had the final piece of the puzzle available to enable them to compute detailed physical models for the Sun and the stars.

References

Bethe, H. A., & Critchfield, C. L. 1938, PhRv, 54, 862

Bethe, H. A. 1939, PhRv, 55, 434; see also
　　Bethe, H. A. 1991, The Road from Los Alamos (New York: American Institute of Physics) 246

Chandrasekhar, S. 1939, An Introduction to the Study of Stellar Structure (Chicago, IL: Univ. Chicago Press) 484

Dalrymple, B. G. 2004, Ancient Earth, Ancient Skies: The Age of the Earth and Its Cosmic Surroundings (Stanford, CA: Stanford University Press)

De Broglie, J. 1926, JPhR, 7, 1

Eddington, A. S. 1926, The Internal Constitution of the Stars (Cambridge: Cambridge Univ. Press); reissued 1988.

Gamow, G. 1928, ZPhy, 51, 200

Gamow, G. 1945, The Birth and Death of the Sun (New York: Penguin Books)

Gurney, R. W., & Condon, E. U. 1928, Natur, 122, 439

Pannekoek, A. 1961, A History of Astronomy (London: George Allen & Unwin Ltd.) 397

---

[4] In the twentieth century, dating methods based on the decays of various radioactive isotopes yielded ages of about 4.5 billion years for the oldest terrestrial rocks, a variety of different meteorites, and—after the Apollo missions to the Moon in the late 1960s and early 1970s—the oldest lunar rocks. The accepted age of the Earth, the Sun, and the rest of the solar system today is 4.54 billion years, cf. Dalrymple (2004).

Chapter 6

Properties of the Matter Inside Stars

In 1939, just as Bethe published his calculations of the rates of thermonuclear reactions in stars, Nazi Germany invaded Poland, igniting World War II. Scientists were among those drawn into the war effort to defeat the Axis Powers. Bethe was tapped to head the Theoretical Division at the top-secret research laboratory constructed at Los Alamos, New Mexico, where an all-out effort got underway to develop the atomic bomb.

Because temperatures at the center of an exploding A-bomb briefly approach those in the core of the Sun, the effort to understand and describe those conditions produced byproducts that were subsequently important in advancing astrophysicists' understanding of stellar interiors. These included greatly improved calculations of opacities, laboratory measurements of nuclear reaction rates, and the beginnings of high-speed numerical calculations of complicated physical processes using electronic digital computers. When the war ended five and a half years later and scientists returned again to take up their peacetime pursuits, they brought with them increased knowledge of the properties of matter under conditions of very high temperatures and pressures, and they applied them to—among other things— improving our understanding of the interiors of the stars.

The specific properties of stellar matter that are needed in order to calculate models for the interiors of the Sun and the stars are—each as functions of the local temperature and density—(1) thermodynamic properties such as the pressure and entropy (a measure of the internal heat content), collectively called the "equation of state;" (2) the opacity; and (3) the rate of energy generation by thermonuclear reactions. In the rest of this chapter, we briefly recount the early history of the development of scientists' understanding of these physical properties.

6.1 The Equation of State

J. Homer Lane's early computations of numerical models for the Sun, subsequent work on polytropes, and Arthur S. Eddington's calculations for stellar interiors all

6-1

assumed that the matter of which the stars are composed behaves as a mixture of ideal gases. With the development of the kinetic theory of gases[1] in 1859 by the great Scottish physicist James Clerk Maxwell (1831–1879) and its subsequent generalization into the subdiscipline of statistical physics[2] by Ludwig Boltzmann (1844–1906) in Germany and Josiah Willard Gibbs (1839–1903) at Yale University in the United States, the thermodynamic properties of ideal gases could be calculated easily. These theoretical calculations reproduced exactly the experimentally measured relationship between the pressure, temperature, and density, and they also yielded results for the specific heats of gases and gas mixtures in agreement with experiments. The approach also enabled more general results, allowing the treatment of mixtures of different gases and—at temperatures low enough to permit the formation of molecules, as occurs in the outer layers of very cool stars—the effects of the internal vibrations and rotation of molecular species.

The high internal temperatures predicted by the stellar models—central temperatures reaching many tens of millions of degrees Kelvin—meant that most of the matter inside the stars not only is in gaseous form but also is almost completely ionized. Today, we refer to such an overall electrically neutral ionized gas as a "plasma." One might naturally expect the electrostatic Coulomb interactions between the ions and the free electrons in such a plasma to produce significant departures from the ordinary ideal gas laws. However, because of the high temperatures, the thermal energies of the electrons and ions are more than 10 times larger than the Coulomb energies, which accordingly produce negligible effects in the equation of state throughout most of a stellar interior. An (important) exception to this generalization occurs in the outer parts of the Sun and stars, where the thermal energies of the particles fall to levels that are comparable to the energies required to ionize the gases.

At lower temperatures, the classical statistical physics developed by Boltzmann and Gibbs has to be generalized to take into account the effects of quantum physics, developed in the early 20th century to describe the interactions between matter and radiation. Under such conditions, two very different types of statistical distributions become important.

One—called the "Fermi–Dirac distribution," after the brilliant Italian physicist Enrico Fermi (1901–1954) and his equally gifted English colleague Paul A. M. Dirac (1902–1984), who invented it in 1926—describes the behavior of elementary particles like electrons and protons, which obey the Pauli Exclusion Principle[3]: No more than one particle of this type can exist at the same time in any given quantum state. Such

---

[1] Kinetic theory considers a gas as a collection of innumerable atoms—idealized as tiny, hard spheres—that move in random directions with random energies, changing directions and energies when they collide with each other, while conserving the overall energy of the system.

[2] Statistical physics considers the physical properties of macroscopic numbers of particles, rather than dealing with the behavior of individual atoms, ions, or electrons.

[3] Wolfgang Pauli (1900–1958) articulated the Exclusion Principle in 1925.

particles are collectively termed "fermions." At the absolute zero of temperature, they fill up all the lowest-energy states of a system (e.g., a system like a chunk of stellar matter) to a maximum level called the "Fermi energy." Such a system is said to be "degenerate." We shall see later that this behavior has profound consequences for our understanding of some of the later phases in the evolution of stars, when the matter of which they are composed is compressed to very high densities, making the Fermi energy very large.

The second type of distribution—called the "Bose–Einstein distribution" after the Indian mathematical physicist Satyendra N. Bose (1894–1974) and Albert Einstein (1879–1955), who proposed it in 1924—applies to fundamental particles such as photons. These are called "bosons," and an arbitrarily large number of such particles can occupy the same quantum state at the same time.

The simplest equations of state—which, however, suffice to describe the properties of stellar matter in most cases—consider the fundamental constituents of the gases to be non-interacting particles; that is, these equations of state neglect the forces acting between particles. Statistical physics is also capable in principle of treating systems of particles that do interact with each other, though in practical calculations, the inter-particle interactions are generally treated in some form of approximation, and the more detailed the treatment, the more complicated the calculations become. Such calculations are generally necessary only in dealing with increasingly involved computations of opacities, but they are not relevant to this part of our story.

6.2 Opacity Calculations

As Eddington had pointed out in the early 20th century, numerical models for stellar interiors that include radiative zones (regions through which energy is transported by the flow of radiation) require knowledge of the opacity of the stellar matter. As noted previously, the processes that contribute to the opacity in stellar interiors are (1) the scattering of photons by electrons and (2) the absorption of photons by free–free, bound–free, and bound–bound transitions in ions and atoms.

The first of these processes does not involve absorption. Instead, scattering involves the collision of a photon with an electron, resulting in a redirection of the momenta of both quantum particles. This produces a contribution to the total opacity that is independent of the photon frequency.

The so-called "free–free" contribution to the opacity is the absorption of a photon by an unbound electron passing near a positive ion, which raises the electron into another unbound quantum level with higher energy. This contribution to the opacity increases as the square of the ionic charge and declines with the cube of the photon frequency. Thus, an ultraviolet photon with a wavelength of 3600 Å is absorbed by the free–free process eight times less efficiently than a photon in the red portion of the spectrum with a wavelength of 7200 Å, which has half the energy of the ultraviolet photon. There are contributions to the free–free opacity from every ionization stage of every element, and the contribution from each is proportional to the relative abundance of that species. It is accordingly necessary to calculate the fractions of each chemical element in each possible ionization state in order to compute this contribution to the total opacity.[4]

In the bound–free absorption process—also called "photoionization"—an electron in a bound state of an atom or ion absorbs a photon with sufficient energy to eject the electron into an unbound state. Photons with energies less than the threshold for ionization from a particular quantum state of a particular atom or ion—called the "absorption edge" for that species—are not affected at all by this process, while photons with energies just above the threshold are strongly absorbed. As with free–free absorption, the efficiency of bound–free absorption also drops off as the cube of the photon frequency. Every internal energy state of every ionization stage of every element contributes to the bound–free opacity, so in principle there may be thousands of contributions to be calculated for the bound–free opacity of a mixture of gases.

The final contribution to the opacity results from the absorption of a photon by an electron in one internal quantum state of an atom or ion that raises it to a higher bound level. This is called bound–bound—or "line"—absorption. While this process is confined to a relatively narrow range of frequencies associated with each absorption line, there may be millions of line transitions to be taken into account.

Once all of the frequency-dependent contributions to the opacity of stellar matter have been obtained, the final calculation to be done is the computation of the "Rosseland mean" opacity, which is the appropriate average that governs radiative energy flow through the interior of a star (Rosseland 1924).

In 1932, Danish astrophysicist Bengt Strömgren (1908–1987) carried out the first quantitative calculations of the Rosseland mean opacities of stellar matter (Strömgren 1932), assuming a composition of hydrogen, helium, and a "Russell mixture" of heavier elements.[5] At the time of this work, the solar abundances of the lightest two elements had not been established, so Strömgren's calculations focused on hydrogen mass fractions about half the value accepted today. Further, at the high temperatures he considered—appropriate for the deep interiors of stars but *not* to regions closer to the surface than about 20% of the solar radius—hydrogen and helium are fully ionized and so do not contribute significantly to the opacity.

Eight years later, MIT's Philip Morse (1903–1985) calculated improved values for the radiative opacities (Morse 1940), using improved quantum physics techniques to extend Strömgren's results to a wider range of temperatures and densities. Morse's results encompassed compositions including hydrogen and helium in any proportions, accompanied by one of four different mixtures of heavy elements. Like

---

[4] Calculations of the statistical populations of each ionization state in a gas in thermal equilibrium were first obtained by the Indian astrophysicist Meghnad Saha (1892–1956) in 1921, and the corresponding equation bears his name. See Saha, M. N. 1921, ZPhy, **6**, 40, "Versuch einer Theorie der physikalischen Erscheinungen bei hohen Temperaturen mit Anwendungen auf die Astrophysik."

[5] A "Russell mixture" is a mixture of heavy elements in proportions first measured for the solar atmosphere by Princeton's Henry Norris Russell in 1929. These proportions are approximately 4/8 by mass of Na plus Mg; 1/8 of Si; 1/8 of K plus Ca; and 2/8 of Fe. Strömgren also included an amount of oxygen equal in mass to all of these metallic elements combined.

Strömgren, Morse included contributions to the opacity from bound–free, free–free, and scattering processes, but not from bound–bound (line) transitions.

A decade after the end of World War II, Peter Naur remarked that (Naur 1954) "There exist, at present, no detailed computations of the opacity of mixtures of elements which include relatively large amounts of hydrogen and helium." The situation improved during the following year, when Geoffrey Keller and Roland E. Meyerott, who had been working at Argonne National Laboratory with the support of the U.S. Atomic Energy Commission, published tables of opacities for a number of gas mixtures (Keller & Meyerott 1955). Their calculations also did not include bound–bound transitions, although they did include some additional physical effects in the equation of state—to account for the close spacings among atoms and ions at lower temperatures—that influence the opacities. They computed opacities for 13 different mixtures, most of which were rich in hydrogen and helium. While computations of the positions and heights of the absorption edges were still done by hand, the results were then entered onto IBM punched cards, and the calculation of the Rosseland mean opacity was done "on an IBM card-programmed calculator" (Keller & Meyerott 1955). The authors expressed their hope that the calculations might be accurate to about 10%.

The detonation of the first hydrogen bomb in 1952 (powered by fusion reactions like those that sustain the Sun) focused the attention of the U.S weapons laboratories on physical conditions that bear an even stronger resemblance to the interiors of the stars than does the interior of an exploding atomic bomb (powered by the fission of heavy elements like uranium and plutonium). In particular, in order to calculate accurately the manner in which the explosion of an H-bomb develops, it became necessary to compute even more-accurate opacities than had been available up to that point. As Arthur N. Cox (1927–2013) put it a dozen years later, "For the past 10 years, some of us at the Los Alamos Scientific Laboratory have been interested in opacities for a wide variety of purposes including astrophysical problems" (Cox 1964).

Making good use of the electronic digital computers that were being installed at the weapons laboratories, as well as at leading research universities, during the late 1950s and early 60s, Cox and his Los Alamos colleague John N. Stewart in 1965 published a new set of calculations of opacities for use in astrophysics (Cox & Stewart 1965). An example of these calculations is shown in Figure 6.1, for the mixture Cox and Stewart termed "Cameron I" after nuclear astrophysicist Alastair G. W. Cameron (1925–2005), who requested results for a particular set of element abundances with $X = 0.74$ and $Y = 0.24$, similar to current values of the hydrogen and helium abundances, respectively. The large peak in the opacity near a temperature of 20,000 K is due to absorption processes accompanying the ionization of hydrogen, while the gentle bump in the descending curve near $\log T = 5$ is associated with the thermal ionization of helium. As the authors noted, this 1965 publication was "the result of numerous private requests for opacities" (Cox & Stewart 1965), and these were the first opacity calculations to include bound–bound contributions in addition to the contributions considered by earlier workers. Continued

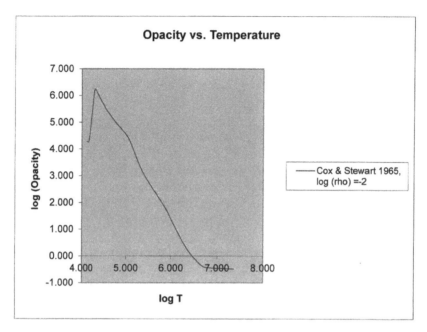

Figure 6.1. The logarithm of the Rosseland mean opacity for the mixture "Cameron I" as a function of the logarithm of the temperature in Kelvins at the density 10^{-2} gm·cm^{-3}. From Cox and Stewart (1965) as plotted by the author. Starting at low temperatures (log T = 4.0, or T = 10,000 K), the opacity first increases with increasing temperature as the statistical populations of the excited states increase. Beyond the peak at about 20,000 K (log $T \approx$ 4.3), more and more of the atoms become ionized, and the opacity declines as the temperature increases further. Above about 5×10^6 K (log $T \approx$ 6.7), essentially the only opacity source that remains is the scattering of photons by free electrons.

refinements and extensions of the work by Cox and his collaborators were collected together in the Los Alamos Opacity Library.

The Los Alamos calculations became the "gold standard" for opacities throughout the 1960s and 70s. They enabled astrophysicists throughout the world to achieve major advances in understanding the structure and evolution of the Sun and the stars. Further important advances in opacity calculations occurred during the 1980s and 90s, but that is a later part of the story.

6.3 Thermonuclear Reaction Rates

In the years following World War II, scientists at Caltech's Kellogg Radiation Laboratory undertook measurements of nuclear reaction rates for a variety of elements of astrophysical importance. Built in 1931 with funding from cereal magnate W. K. Kellogg, the laboratory was established under the directorship of Danish-born physicist Charles Christian Lauritsen (1892–1968), with the dual mission of treating cancer patients and studying the interactions of protons, deuterons, and alpha particles with light nuclei (Greenberg & Goodstein 1983a). The cancer treatments ended just before the start of World War II, but Lauritsen continued the nuclear physics research. This positioned the laboratory perfectly to test the various reactions involved in the CNO bi-cycle when scientists returned at the end of the war. As Kellogg physicist and subsequent Nobel laureate William A. Fowler (1911–1995) commented in 1983, "When Bethe came out with the carbon–nitrogen cycle, we … felt a proprietary interest in this group of reactions because we had been working on them…" (Greenberg & Goodstein 1983b) Consequently, in 1946 Kellogg began its low-energy nuclear astrophysics research, and in 1950 Fowler and one of his doctoral students published their first results (Hall & Fowler 1950). This experimental work at Caltech during the next several decades provided fundamental information about the rates of hundreds of nuclear reactions that were essential to achieving an understanding not only of stars along the Main Sequence but also of advanced stages in the evolution of stars. They also provided the foundation for our understanding of the origins of the chemical elements. We shall return to both matters later in this book.

And in 1952, Bethe's younger colleague at Cornell, Edwin E. Salpeter (1924–2008) published careful re-calculations of the rates of reactions involved in the proton–proton chain (Salpeter 1952), using the most up-to-date information then available. In contrast to the CNO cycle, for which rates can actually be measured in the laboratory, the proton–proton reaction proceeds much too slowly to enable laboratory measurements and must therefore by necessity be calculated theoretically.

References

Cox, A. N. 1964, JQSRT, 4, 737

Cox, A. N., & Stewart, J. N. 1965, ApJS, 11, 22

Greenberg, J. L., & Goodstein, J. R. 1983a, Paper presented at the History of Science Annual Meeting, Norwalk, 29 October

Greenberg, J. L., & Goodstein, J. R. 1983b, as "William A. Fowler," transcript of an oral history interview conducted by Charles Weiner, American Institute of Physics, June 8 and 9, 1972, 39

Hall, R. N., & Fowler, W. A. 1950, PhRv, 77, 197

Keller, G., & Meyerott, R. E. 1955, ApJ, 122, 32

Morse, P. M. 1940, ApJ, 92, 27

Naur, P. 1954, ApJ, 119, 365

Rosseland, S. 1924, MNRAS, 84, 525

Salpeter, E. E. 1952, PhRv, 88, 547

Strömgren, B. 1932, ZPC, 4, 118

Part II

A Closer Look at the Solar Interior

Before we go on to consider other stars, in the next few chapters we discuss the present understanding of the inside of the Sun, as revealed by observations made possible by modern instrumentation, including data from spacecraft and elementary-particle detectors.

Chapter 7

How Our Understanding of the Present Sun Developed

The first attempt to construct a complete numerical model for the Sun—including energy generation by thermonuclear reactions—was carried out by Bethe and his collaborators in 1941 (Blanch et al. 1941). They included energy generation only by the extremely temperature-sensitive CNO cycle. However, "the carbon cycle (for a 1% concentration of carbon and nitrogen …) leads to a predicted total luminosity for the Sun equal to 5.6×10^{35} ergs s^{-1}—a value about one hundred times as large as the observed value" (Blanch et al. 1941). After the hiatus caused by World War II, a number of astrophysicists began to consider ways to improve numerical models for the Sun in an effort to resolve the large discrepancy between the observed value of the solar luminosity and that predicted by the 1941 model constructed by Bethe and his colleagues. The problem to be solved, as sketched by astrophysicist Martin Schwarzschild (1912–1997; see Figure 7.1), was this: Mathematically, a stellar interior is described by "four first order differential equations in just one variable, the distance from the center of the star…, with the distance starting out at zero at the center and going out to the radius of the star. The problem has boundary conditions, two in the center and two at the surface… But the equations are extremely non-linear…. [The process of carrying out the numerical calculations] was incredibly tedious and slow. To get a solution for a particular star for one particular time in the life of the star would take two or three months. And if it was a tricky case, a more evolved case, it could take half a year" (Aspray 1986a). The calculations were done on desk calculators using a numerical procedure that required fitting together solutions started from the center and from the surface at some intermediate matching point. "It was a very awkward procedure that got worse and worse as we followed a star in its evolution….," Schwarzschild later recalled. "So there was no question in my mind that if there were better computational tools I certainly wanted to learn them and hoped to apply them" (Aspray 1986b).

Figure 7.1. Martin Schwarzschild. Credit: Emilio Segré Visual Archives; copyright American Institute of Physics.

The son of the eminent German astrophysicist Karl Schwarzschild (1873–1916), Martin was born in 1912 in Potsdam, Germany, where his father was then serving as the director of the observatory. The elder Schwarzschild died in 1916 of illness contracted while fighting in the trenches for the Kaiser during World War I. After his death, the Schwarzschild family moved back to Göttingen, their original hometown. In 1935, Martin received his doctorate in astrophysics from the Universität Göttingen. He left Germany in 1936, along with numerous others of Jewish ancestry who were being persecuted by the Nazis (Trimble 1997).

Schwarzschild first spent a year as a postdoc with Svein Rosseland in Norway. Moving next to the United States, he spent the next three years on a postdoctoral fellowship with Harlow Shapley at the Harvard College Observatory. In 1940, he joined the faculty at Columbia University in New York, where he gained experience with advanced electromechanical IBM machines at the Watson Scientific Computing Laboratory. After Pearl Harbor was bombed in 1941, however,

Schwarzschild joined the U.S. Army as a private, and he became a U.S. citizen in 1942. In January 1943, he was initially assigned to the Ballistics Research Laboratory at Aberdeen, Maryland. He was subsequently ordered to an Army intelligence unit in Italy and was awarded the Legion of Merit and the Bronze Star for his wartime activities. After the war, he married astronomer Barbara Cherry, and in 1947, after Henry Norris Russell retired, Princeton hired both Schwarzschild and astronomer Lyman Spitzer, Jr. (1914–1997).

In 1946, Schwarzschild tackled the problem of the internal structure of the Sun in an effort to improve agreement with the measured solar luminosity. "I was already interested in the theory of the stellar interior and continued to be so...," Schwarzschild later commented (Aspray 1986c). He included helium, as well as hydrogen and a Russell mixture of heavier elements, in the assumed solar composition (Schwarzschild 1946). Like others working to construct numerical solar models around this time, Schwarzschild assumed a convective core overlain by a radiative envelope, with energy generation again assumed to be provided solely by the carbon cycle confined to the convective core. Using improved opacities that had been calculated in 1940 by Philip Morse (Morse 1940), Schwarzschild computed more accurate, new models for the radiative envelope. By fitting his new radiative envelope models to the presumed convective core, he was able to determine the values of the hydrogen and helium mass fractions, X and Y, respectively. Requiring the mass, radius, and luminosity of the model to match the observed values for the Sun yielded $X = 0.47$, $Y = 0.41$, and gave a central temperature of about 20 million Kelvin. Schwarzschild triumphantly concluded "that the introduction of an appreciable helium content removes the discrepancy, found earlier, between the observed luminosity and the rate of energy generation by the carbon cycle" (Schwarzschild 1946).

Schwarzschild's result was not the end of the story, however. About 50 miles northeast of Princeton, astrophysicists at Columbia University in New York City were also working on the problem of the internal structure of the Sun. In a 1948 paper (Keller 1948) based on his doctoral thesis, Geoffrey Keller computed a new solar model with a radiative envelope and a convective core. He assumed the radiative opacity to be due to oxygen, based on a then-recently determined higher abundance of this element. Using Schwarzschild's method to construct his solar model, Keller found similar results for the central temperature, but—as a result of the very different opacities he employed—Keller found quite different results for the hydrogen and helium abundances: $X = 0.67$ and $Y = 0.29$, much closer to the currently accepted values than the nearly equal abundances Schwarzschild had obtained.

Three years later, Schwarzschild's former Ph.D. student, Estonia-born astrophysicist Isadore Epstein (1919–1995), also at Columbia, became the first to include energy generation by both the proton–proton chain and the CN cycle in a solar model calculation, and he questioned the validity of the assumption that the core of the Sun was convective. The results of Epstein's model calculations gave a central temperature of about 15 million Kelvin, with the p–p chain producing 95% of the energy, and the abundances for his solar model were $X = 0.82$ and $Y = 0.17$. In 1953,

Figure 7.2. Erika Böhm–Vitense. Source: Department of Physics, University of Illinois at Urbana–Champaign, courtesy of AIP Emilio Segré Visual Archives; copyright American Institute of Physics.

Epstein and Lloyd Motz (1909–2004) revised this model, employing a then-new calculation of the rate of the proton–proton reaction (Salpeter 1952). They found it possible to fit the mass, radius, and luminosity of the Sun, but only with the helium abundance set to zero, giving $X = 0.998$. That is, their model for the Sun was almost completely pure hydrogen!

In 1953, the young German astrophysicist Erika Vitense[1] (1923–2017; Figure 7.2) demonstrated for the first time that energy transport through the outer layers of the solar interior occurs by means of convection rather than by radiation (Vitense 1953), as had been assumed up to then for all the post-war solar models. In 1906, Martin Schwarzschild's father, Karl, had demonstrated that convection develops spontaneously in a fluid in hydrostatic equilibrium under gravity when the local gradient of the temperature becomes steeper than the so-called "adiabatic" temperature

---

[1] After marrying astrophysicist Karl–Heinz Böhm in 1953, she became Erika Böhm–Vitense.

gradient, the specific temperature gradient at which a parcel of matter displaced from its initial location neither gains nor loses heat to its surroundings. Vitense pointed out that in the region below the solar surface where hydrogen becomes progressively ionized as the local temperature increases inward, two effects occur that cause the hydrogen partial-ionization zone to become convectively unstable. First, as hydrogen begins to become ionized, the opacity increases greatly, as shown in Figure 6.2, and this causes the temperature gradient that would result from radiative transfer to become much steeper; that is, the higher opacity makes radiative transfer much less efficient. The second effect is that in the ionizing gas, energy goes into freeing electrons from the ions rather than into heating the gas. This depresses the adiabatic temperature gradient, making it increasingly prone to convective instability. In combination, these two effects mean that in the hydrogen-ionization zone below the solar surface, energy is transported by convection rather than by radiation.

Meanwhile back in Princeton, Martin Schwarzschild had attracted a cadre of young colleagues who shared his interest in constructing detailed numerical models for the interiors of the stars. By that time, the science associated with stellar interiors had advanced to the point where, for the first time, there was great excitement and enthusiasm in the astrophysical community about the prospects for achieving a genuine understanding of the structure and evolution of the stars. About this time, the gifted mathematician John von Neumann (1903–1957), who had been heavily involved with the Manhattan Project and the development of electronic digital computers during World War II, arranged to have a digital computer—dubbed the "MANIAC"—constructed at Princeton. This was not an unalloyed blessing, however, as the machine frequently broke down. As Schwarzschild later recalled, the MANIAC was built with "all tubes, not yet transistors. If something went wrong and they identified which tube it was, then to get at that tube usually meant cutting a whole slew of wires, not just the ones to the tube. Just to get at the tube you had to cut dozens of wires, keep track of them, get at the tube, replace the tube, and resolder the wires. It was a major undertaking whenever a tube went bad, particularly those that were in the main body of the computer. So the down time was high in part because the failure rate was still very high and in part because the repair time was very high. Also, the main memory consisted of a wide belt magnetic tape, but in a drum form, something on the order of a foot or two long and a foot or two wide and the adjustment of the reading and tracking heads on that drive were extremely delicate and quite often failed (Aspray 1986d)." However, "It just happened that these stellar interior problems were exactly of the kind that was important for von Neumann... That was why he encouraged me to work directly with Herman Goldstine[2] and then later, very extensively with Heddie Selberg, who was one of

---

[2] A lieutenant in the U.S. Army during World War II, Herman Goldstine (1913–2004) had helped to secure funding from the Army in 1943 for the construction of the Electronic Numerical Integrator and Computer (called "ENIAC") at the University of Pennsylvania.

the members of this group, to get the equations onto the MANIAC and start using the MANIAC" (Aspray 1986c). This experience gave Schwarzschild "a marvelous chance to learn how to think in terms of modern computers with von Neumann's logic and to get into that way of approaching numerical problems; so that as the more serviceable machines, particularly the commercial ones, became available I was wonderfully trained to use them" (Aspray 1986e).

The proliferation of increasingly capable digital computers during the 1950s and early 1960s for the first time held the promise of making possible the rapid execution of the enormous numbers of detailed calculations necessary to construct a single stellar model, enabling astrophysicists to begin exploring the quantitative results of differing assumptions about the internal physical properties of the stars. Schwarzschild recalled the excitement of the times: "By 1938 with the famous paper by Bethe and independently by von Weizsäcker the last missing ... major physics in the stellar interior had become clear.... Suddenly [with the advent of electronic digital computers] you had the possibility of computing evolution sequences for individual stars, of comparing different stars and different phases of the evolution with observed stars. There was an enormous wealth already of observations relevant to this question. Suddenly you could start to interpret the observed stars in terms of different masses, different compositions, and different ages. So a wealth (what you might call a zoo) of observed stars fell into patterns and we could start getting ages for stars, particularly for groups of stars in clusters which had common compositions, and probably even common ages. Age determinations were much more secure for them than for individual stars. That gave us the first, however inaccurate age determinations of individual objects in the astronomical scene. Since then, it has evolved into a much more multi-faceted field particularly when ... the white dwarves, which had been understood before but could now be understood as end phases of the lighter stars, and then the neutron stars and finally at least in theory the black holes could all be by and by incorporated.... I think it is not exaggerated to say that the theory of the stellar interior and stellar evolution with all these modern facets is a significant part of modern astrophysics ..." (Aspray 1986f).

In 1954, Peter Naur, a Danish-born computer scientist then working on his doctorate at Yerkes Observatory, pointed out that, in their calculations published the previous year, Epstein and Motz had not included the free–free opacity due to hydrogen in their calculations. Naur computed two additional models for the Sun, each separated into an interior region and an exterior region, the two regions having different assumed opacity laws (Naur 1954). Naur's Model I was unusual for the time in having no convective core. Both models had central temperatures near 14 million Kelvin, central densities of about 90 g cm^{-3}, and hydrogen mass fractions of about $X = 0.75$.

Three years later, Schwarzschild and his colleagues again took up the problem of the solar interior, for the first time "substituting convective equilibrium for radiative equilibrium in the outer parts of the envelope [the necessity of which Erika Vitense had demonstrated] and ... taking into account the inhomogeneities caused by the hydrogen burning during the past life of the Sun" (Schwarzschild et al. 1957).

They began by constructing an initial, homogeneous model for the Sun at the stage in which hydrogen burning nuclear reactions have just begun, assuming it to consist of a radiative interior and a convective envelope. From this model, they computed the rate of transmutation of hydrogen into helium and—assuming the Sun to have been burning hydrogen for five billion years via the proton–proton reactions—they computed the present composition at every point in the solar interior. Finally, based on the new distributions of hydrogen and helium, they constructed a model for the present Sun.

These calculations led to several important results. First, the addition of the convective envelope—which they found to extend over the outermost 18% of the solar radius—allowed the authors to construct a model with heavy-element abundances that agreed with the spectroscopic data. Second, they found that in the innermost 10% or so of the solar mass, nuclear burning over the past five billion years had substantially reduced the mass fraction of hydrogen—to about $X = 0.3$—and correspondingly increased the mass fraction of helium. The authors considered three possible models for the Sun, with initial hydrogen mass fractions of 0.6, 0.7, and 0.8. For $X = 0.7$, which is reasonably close to the modern value, Schwarzschild and his colleagues found a helium abundance $Y = 0.26$. They also found the temperature and density at the Sun's center to be 15.8 million K and 127 g cm^{-3}, respectively. The authors acknowledge that a further improvement in the models would result from including energy generation produced by the CNO bi-cycle reactions as well as that produced by the proton–proton chain. An interesting and unexpected outcome of these calculations was the finding that the luminosity of the Sun has increased by about 60% over the past five billion years.

It is important to note, however, that these calculations are not stellar evolution calculations as we understand them today, but they did for the first time include all of the key elements necessary. Although the calculations were carried out using the electronic digital computer at Princeton's Institute for Advanced Study, Schwarzschild and his colleagues still employed the fitting technique that had been used since the early 20th century.

Meanwhile, out on the West Coast, astrophysicists at Berkeley were developing an entirely new approach to stellar-structure calculations, using the electronic digital computers that were just starting to become generally available during the 1950s. Astrophysicist Louis G. Henyey (1910–1970; see Bodenheimer 1995) for a brief biography and photograph of Henyey) realized that if a stellar model were assumed always to be in hydrostatic equilibrium, the minuscule time steps required for fully hydrodynamic calculations would not be necessary.[3] Instead, calculations of stellar evolution could be carried out using the much longer time

---

[3] The hydrodynamic timescale for the Sun, for example, is about half an hour—the time it takes for a sound wave to travel across the solar radius. It would have far exceeded the capabilities of early digital computers to follow perhaps millions or billions of years of evolution for even one star in time steps this short!

steps that are sufficient to follow changes in the heat content of different parts of the star. These are the interesting timescales for stellar evolution—those accompanying changes in the thermonuclear energy-production rates and the chemical composition of the star, variations in heat flow accompanied by the growth or shrinkage of stellar convection zones and radiative regions, and the changes accompanying gravitational expansion or contraction. Henyey and his colleagues embodied this new insight into a computer code that they used to follow the gravitational contraction of stars without nuclear energy sources as they approached the Main Sequence. The "Henyey method," as it came to be called, would soon sweep the world and revolutionize our understanding of stellar structure and evolution.

As Schwarzschild later recalled this development, "...when the hydrogen bomb development came, Johnny Wheeler here at the university [in Princeton] had a project of deriving theoretical conclusions about how the detonation in a hydrogen bomb might occur. He was joined for one year by a theoretical astronomer from Berkeley, Louis Henyey. I was not involved. I had not volunteered for that work and did not have the proper clearances. That was my choice.... After Henyey returned to Berkeley after this year, he realized what he had learned here was extremely useful for the stellar interior. He applied a general scheme that ... [had been] developed specifically for boundary value problems. And Henyey wrote one fundamental paper about that method applied to the stellar interior. We used the term "Henyey method" for it, but the basic idea he learned here... either directly or indirectly from von Neumann. And the moment Henyey's paper came out in the open ... astronomical literature, and after I understood it, ... I have never used anything else and it became the standard tool... for the theory of the stellar interior" (Aspray 1986c).

The method Henyey and his colleagues developed for computing time sequences of stellar models automatically using electronic digital computers (Henyey et al. 1959) was an outgrowth of a method originally devised by Sir Isaac Newton three centuries earlier. In brief, the method starts with an approximate model for the distributions of pressure, temperature, etc., within a star of given total mass. Since the approximate model does not in general satisfy the finite-difference equations that approximate the differential equations for numerical calculations, the computer program computes corrections to all the physical quantities at each mass shell within the model. Adding the computed corrections to each of the physical quantities yields an improved approximation to the model. The process of successively improving the model quantities is continued until the computed corrections become smaller than some predetermined level of accuracy, when the model is said to have converged. The converged model is then used as the initial approximation for a new model at a slightly later point in time, and the iterative process of computing successive corrections to the physical variables is repeated. In this way, a sequence of models representing successive steps in the evolution of the star can be computed rapidly.

Henyey et al. encountered two problems in their initial applications of this procedure. First, the arbitrariness of the starting model raised some initial concerns about the validity of the final results. However, by trying several different types of starting models, the investigators found that differences in the calculations gradually decayed away as the model was evolved. Second, they experienced some initial difficulties in dealing with the switch between radiative and convective energy transport within a given model. This problem was overcome by an adjustment in the computer program (Henyey et al. 1964).

The new "Henyey method" completely transformed the calculation of models for the structures of the Sun and the stars. What had previously required months of mind-numbing drudgery could now be done automatically in a few seconds on a digital computer. The ease of model construction also meant that astrophysicists could now employ much more complicated approximations for the equation of state, opacity, and nuclear reaction rates, making possible the construction of increasingly realistic models. It also meant that astrophysicists could construct not just individual static models but instead could compute entire sequences of models representing the evolution of the internal structures of the stars through time. We shall see examples of such calculations later in this book. For now, however, our focus remains on the Sun.

It did not take long for astrophysicists to explore the evolution of the Sun, from an initial phase of gravitational contraction, leading to the heating of the interior, to the ignition of hydrogen burning in the deep interior as the Sun settled onto the Main Sequence, and through the changes resulting from 4.5 billion years of evolution. Thus, by the end of the 1960s, astrophysicists had finally achieved a realistic understanding of the internal structure of the present Sun. Of course, improvements continued to be made—and still continue to the present day—but they increasingly take the form of fine tuning a well-established model.

To select just two examples, let us consider a series of solar models constructed using a variety of different modern approximations for the material properties of the solar interior by the Yale group (Guenther et al. 1992) and by the Princeton group (Bahcall et al. 2005). Among the ten models in the former collection and the seven in the latter, the spread in the calculated values of the model parameters is impressively small. For example, the value of the initial helium abundances determined from the models is $Y = 0.28 \pm 0.01$; the logarithms of the central temperature and density, respectively, are 7.189 ± 0.004 (corresponding to a temperature of 15.5 million Kelvin) and 2.169 ± 0.007 (corresponding to 148 g cm^{-3}); and the bottom of the convection zone in the solar envelope lies at a fraction 0.73 ± 0.01 of the solar radius (that is, the convection zone extends downward from the solar surface 27% of the distance to the Sun's center); see Figure 7.3.

While it is gratifying to find such close agreement among so many of the best current models for the Sun, what assurance do we really have that they actually do represent the true conditions in the solar interior? Is there any way to test the models directly? The next two chapters address these questions.

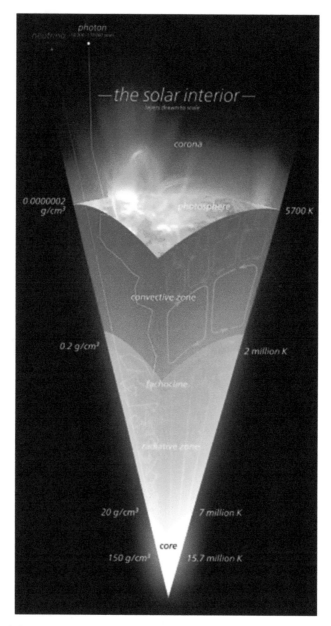

Figure 7.3. An artist's conception of the internal structure of the present Sun. The decrease in temperature from 15.7 million K at the center to 5700 K at the solar photosphere is indicated along the right-hand edge of this pyramidal core. The corresponding decrease in the density is shown along the left-hand edge. Energy is transported entirely by radiation through the innermost parts of the solar interior. The subsurface convection zone—colored red in this image—carries the energy from the transition region on out to the solar photosphere (the visible surface of the Sun), where it is radiated away into space as sunlight. The tops of the convection cells are visible as the "solar granulation" at the Sun's surface, and the solar corona—with magnetically confined loops of highly ionized gases—extends on out to produce the solar wind that is constantly blowing outward from the Sun. Source: commons.wikimedia.org/wiki/File:The_solar_interior.svg.

References

Aspray, W. 1986a, An Interview with Martin Schwarzschild, OH 124 (Minneapolis, MN: Charles Babbage Institute, Univ. Minnesota) 12

Aspray, W. 1986b, An Interview with Martin Schwarzschild, OH 124 (Minneapolis, MN: Charles Babbage Institute, Univ. Minnesota) 13

Aspray, W. 1986c, An Interview with Martin Schwarzschild, OH 124 (Minneapolis, MN: Charles Babbage Institute, Univ. Minnesota) 15

Aspray, W. 1986d, An Interview with Martin Schwarzschild, OH 124 (Minneapolis, MN: Charles Babbage Institute, Univ. Minnesota) 20

Aspray, W. 1986e, An Interview with Martin Schwarzschild, OH 124 (Minneapolis, MN: Charles Babbage Institute, Univ. Minnesota) 14

Aspray, W. 1986f, An Interview with Martin Schwarzschild, OH 124 (Minneapolis, MN: Charles Babbage Institute, Univ. Minnesota) 18

Bahcall, J. N., Serenelli, A. M., & Basu, S. 2005, ApJ, 621, L85

Blanch, G., Lowan, A. N., Marshak, R. E., & Bethe, H. A. 1941, ApJ, 94, 37

Bodenheimer, P. H. 1995, Biographical Memoirs (Vol. 66; Washington, DC: National Academies Press) 166

Guenther, D. B., Demarque, P., Kim, Y.-C., & Pinsonneault, M. H. 1992, ApJ, 387, 372

Henyey, L. G., Forbes, J. E., & Gould, N. L. 1964, ApJ, 139, 306

Henyey, L. G., Wilets, L., Böhm, K. H., LeLevier, R., & Levee, R. D. 1959, ApJ, 129, 628; see also

Henyey, L. G., LeLevier, R., & Levée, R. D. 1955, PASP, 67, 154; and

Henyey, L. G., LeLevier, R., & Levée, R. D. 1959, ApJ, 129, 2

Keller, G. 1948, ApJ, 108, 347

Morse, P. M. 1940, ApJ, 92, 27

Naur, P. 1954, ApJ, 119, 365

Salpeter, E. E. 1952, PhRv, 88, 547

Schwarzschild, M. 1946, ApJ, 104, 203

Schwarzschild, M., Howard, R., & Härm, R. 1957, ApJ, 125, 233

Trimble, V. 1997, PASP, 109, 1289

Vitense, E. 1953, ZA, 32, 135

Chapter 8

Ghost Particles from the Center of the Sun

Several of the nuclear reactions involved in hydrogen burning emit subatomic particles called "neutrinos." Each cycle through the CNO reactions releases two; one neutrino from the positron[1] decay of ^{13}N and the second from the decay of ^{15}O. In addition, three reactions in the proton–proton chain emit neutrinos: the basic proton–proton reaction,

$$p + p \rightarrow d + e^+ + v_e,$$

where the symbol "d" represents the deuteron—a proton and a neutron bound together by the strong nuclear force—"e^+" represents a positron, and "v_e" is the electron neutrino; the proton–electron–proton, or "pep," reaction,

$$p + e^- + p \rightarrow d + v_e;$$

and the positron decay of a radioactive isotope produced in one of the branches of the proton–proton chain,

$$^8B \rightarrow {}^8Be^* + e^+ + v_e.$$

Here $^8Be^*$ is the unstable beryllium-8 nucleus, which breaks up into two alpha particles (^{4}He nuclei) almost as soon as it is created. As we shall see, the neutrinos emitted by reactions in the proton–proton chain are the ones to which terrestrial neutrino detectors are most sensitive.

Neutrinos interact so weakly with matter that they escape from the center of the Sun as readily as visible light passes through a transparent windowpane. Accordingly, if we can detect and measure the flux of neutrinos from the Sun, we can obtain a direct measurement of the rates of nuclear reactions going on *today*

---

[1] The positron is the antiparticle of the electron. It has the same mass but possesses a positive, rather than a negative, electrical charge. When a positron and an electron collide, they completely annihilate each other, producing a burst of gamma radiation (high-energy photons).

right down in the center of our star. This obviously has the potential to provide an important test of the solar models, because—as astrophysicist John N. Bahcall (1934–2005) noted—"The predicted number of high-energy solar neutrinos can be shown ... to depend sensitively on the central temperature of the Sun. A 1% error in the temperature corresponds to about a 30% error in the predicted number of neutrinos; a 3% error in the temperature results in a factor of two error in the neutrinos..." (Bahcall 2004). For this reason, even a fairly rough measurement of the solar neutrino flux can place a rather tight constraint on the temperatures deep inside the Sun.

So, what are neutrinos, and how can we go about measuring the flux of neutrinos produced in the Sun?

The concept of the neutrino was introduced by the great Austrian-born physicist Wolfgang Pauli (1900–1958) during informal comments at a Solvay Conference in Brussels in 1931 (Pauli 1934). He felt compelled to propose the existence of this particle in order to enable physicists to retain the fundamental physical law of energy conservation. Without the neutrino, energy otherwise appeared *not* to be conserved during the process of beta decay. This process—the spontaneous emission of electrons (or positrons) by nuclei as they transform into isotopes of greater (respectively, lesser) atomic number—had been discovered at the end of the 19th century. In 1934, the brilliant Italian physicist Enrico Fermi (1901–1954) constructed a quantum theory of beta decay based on the neutrino hypothesis (Fermi 1934), and it accounted quantitatively for the shape of the electron energy spectrum observed during this process. During the following decades, Fermi's original theory was improved and extended several times.

The ghostly neutrino does not take part in *any* interactions with material particles other than the weak interaction itself—the subatomic force that causes beta decay. "Neutrinos have zero electric charge, interact very rarely with matter, and—according to the textbook version of the standard model of particle physics—are massless. About 100 billion neutrinos from the Sun pass through your thumbnail every second, but you do not feel them because they interact so rarely and weakly with matter... There are three known types of neutrinos. Nuclear fusion in the Sun produces only neutrinos that are associated with electrons, the so-called electron neutrinos (v_e). The two other types of neutrinos, muon neutrinos (v_μ) and tau neutrinos (v_τ) are produced, for example, in laboratory accelerators ... together with heavier versions of the electron, the particles muon ... and tau..." (Bahcall 2004).

Originally, the neutrino was thought to have zero mass like the photon, the quantum of electromagnetic radiation, but now it appears that instead it actually has a very small mass. Looking back in 2004, after the solar neutrino problem had finally been resolved, John Bahcall wrote, "The simplest [modification to the standard model of particle physics] that fits all the neutrino data implies that the mass of the electron neutrino is about 100 million times smaller than the mass of the electron" (Bahcall 2004).

The actual existence of neutrinos was first demonstrated (Cowan et al. 1956) in 1956 by Frederick Reines (1918–1998) and Clyde Cowan, Jr. (1919–1974). They carried out "very beautiful and careful experiments in which the neutrinos from a nuclear reactor ... [were] allowed to fall on the hydrogen of an organic scintillator solution" (Preston 1962). Occasionally, a neutrino interacted with a proton to produce a neutron and a positron, $v_e + p \rightarrow n + e^+$, effectively the inverse of the process of neutron decay. This was quickly followed "by two pulses in the scintillator, one produced by the ... [gamma] radiation [emitted when the positron collides with an ambient electron and is annihilated] and the other by the ... [gamma ray] emitted when the neutron is captured" (Preston 1962). Because neutrinos interact so weakly with matter, "several hundred liters of scintillator [fluid] are required and about a hundred photomultipliers" to detect the optical emission produced by the scintillator fluid.

As the 1950s gave way to the 1960s, the physical existence of the neutrino had thus been demonstrated experimentally, and models for the solar interior had been improved to the point where scientists were reasonably confident of the results. The stage had been set for the next major advance in understanding the inside of the Sun.

The first calculations of the flux of neutrinos from the Sun, based on detailed solar models, were published in 1963 by William A. Fowler, Icko Iben, Jr., Richard L. Sears and John N. Bahcall (Bahcall et al. 1963). Their results indicated that the expected capture rate of solar neutrinos should be about 50 SNU[2]. The following year, in companion papers in the scientific journal *Physical Review Letters*, Bahcall and Brookhaven National Laboratory physical chemist Raymond Davis, Jr. (1914– 2006), proposed an experiment to test the solar models by measuring directly the flux of neutrinos emitted from thermonuclear reactions in the Sun's core.

Born in Shreveport, Louisiana, John N. Bahcall (Figure 8.1) initially aspired to a career as a rabbi (http://en.wikipedia.org/wiki/John_N_Bahcall). He began college as a philosophy major at the Louisiana State University, transferring to the University of California at Berkeley, where he received his A.B. in 1956. A Berkeley physics course he took as a graduation requirement completely transformed his career path, and he received his subsequent degrees in physics (M.S. from the University of Chicago in 1957 and Ph.D. from Harvard in 1961). He met his future wife, Neta, herself an outstanding astrophysicist, during a visit to the Weizmann Institute as a graduate student. After eight years at Caltech—where he began the solar-neutrino work with Ray Davis to which he would devote much of his career—he became a professor at the Institute for Advanced Study in Princeton in 1971. He was "one of the scientific Brahmins of his time, leading numerous influential committees and organizations ... and advising NASA and Congress" (Overbye, 2005).

Raymond Davis, Jr. (Figure 8.2) was born in Washington, DC, where his father worked for the National Bureau of Standards (now called the National Institute for Standards and Technology, NIST) (http://en.wikipedia.org/wiki/Raymond_ Davis_Jr). He received his Bachelor's and Master's degrees in chemistry from the University of Maryland and in 1942 obtained his Ph.D. in physical chemistry from Yale University. After his discharge from the U.S. Army in 1946, Davis "went to

---

[2] The "solar neutrino unit," or "SNU"—pronounced "snew"—equals 10^{-36} solar–neutrino interactions per target nucleus per second.

Figure 8.1. John N. Bahcall in 1976. Credit: AIP Emilio Segré Visual Archives, John Irwin Slide Collection. Copyright: American Institute of Physics; copyright American Institute of Physics.

work at Monsanto's Mound Laboratory, in Miamisburg, Ohio, doing applied radiochemistry ... [for] the United States Atomic Energy Commission (http://en. wikipedia.org/wiki/Raymond_Davis_Jr)." He joined Brookhaven National Laboratory in 1948, remaining there until 1984, when he joined the research faculty of the University of Pennsylvania. "When Davis arrived at Brookhaven in 1948, the existence of neutrinos was still speculative. Looking for a challenging problem that could be addressed with physical chemistry techniques, he decided to develop and construct a chlorine-based neutrino detector that had been described by Bruno Pontecorvo two years earlier.... Both William Fowler of Caltech and Alastair

Figure 8.2. Left-to-right: John Galvin, Raymond Davis, Jr., and Bruce Cleveland (the Solar Neutrino Group at Brookhaven National Laboratory in 1977) in front of a photograph of the original solar-neutrino detector in the Homestake Mine. Credit: Courtesy of Brookhaven National Laboratory; used with permission.

Cameron of Atomic Energy of Canada Ltd urged Davis to use his detector to look for neutrinos produced by fusion in the solar core..." (Lande 2006).

Bahcall's detailed 1964 calculations of the solar-neutrino flux and spectrum employed the best models of the solar interior then available (Bahcall 1964). In his companion paper to Bahcall's theoretical calculations, Davis proposed to detect neutrinos from the Sun by utilizing the reaction $^{37}Cl + v_e \rightarrow {}^{37}Ar + e^-$. The detector was to be a large container of ordinary cleaning fluid (C_2Cl_4) placed deep underground to shield it from cosmic rays, which could otherwise produce false signals (Davis 1964). The ^{37}Ar nucleus is unstable and decays back into ^{37}Cl with a half-life[3] of 35 days. This is long enough to allow the small number of ^{37}Ar atoms produced by the neutrino reaction to be extracted from the tank of fluid by careful chemical processes, and the beta decays of the individual radioactive nuclei could then be detected and counted. As Bahcall pointed out in a 1969 article, only the "pep" reaction and the decay of radioactive ^8B "produce neutrinos energetic enough to trigger a reaction in the tetrachloroethylene detector" (Bahcall 1969).

The initial experiments, using a tank containing 1000 gallons of cleaning fluid in a 2300-foot-deep limestone mine in Ohio, found only an upper limit of less than 0.5 solar neutrinos per day. Although this was only an upper limit, it succeeded in

---

[3] A "half-life" is the time required for exactly half of an initial quantity of radioactive matter to decay.

Figure 8.3. The Brookhaven National Laboratory solar neutrino detector in the Homestake Mine *circa* 1968. The two figures in hardhats provide the scale of the detector. Ray Davis is standing on the catwalk looking down at John Bahcall, standing with his foot on the ladder. Credit: Digital Photo Archive, Department of Energy (DOE), courtesy of AIP Emilio Segré Visual Archives; used with permission.

demonstrating that Davis's experimental approach would work. Based on rates calculated by Bahcall, Davis estimated that a tank containing 100,000 gallons of cleaning fluid placed at a depth of 4500 feet should be capable of unambiguously detecting the expected solar neutrino flux.

These results were sufficiently promising to persuade the administration of Brookhaven National Laboratory to support the construction of large tank detector 20 feet in diameter and 48 feet long placed at a depth of 4850 feet in the Homestake gold mine in Lead, South Dakota (see Figure 8.3). The first Homestake results, announced in 1968 (Davis et al. 1968), showed that the solar neutrino signal was less than 3 SNU, as against a predicted rate of production by solar neutrinos of 8 SNU (Lande 2006). The continuing inability of the chlorine detector to turn up as many neutrinos as predicted by the solar models subsequently prompted one wag to ask, "What SNU?"

The disagreement between Davis's experimental findings and Bahcall's theoretical calculations stimulated efforts both to refine the experiment and to investigate every conceivable problem with the calculations. From the perspective of theory, as Bahcall and his coworkers pointed out (Bahcall et al. 2001), "to predict correctly the number of neutrinos produced by nuclear reactions in the Sun, many complicated phenomena must be understood in detail. For example, one must understand a smorgasbord of nuclear reactions at energies where measurements are difficult. One must understand the transport of energy at very high temperatures and densities. One must understand the state of the solar matter in conditions that cannot be studied directly on Earth.... One must measure the abundances of the heavy elements on the surface of the Sun and then understand how these abundances change as one goes deeper into the Sun. All of these and many more details must be understood and calculated accurately" (Bahcall 2004). Bahcall and his colleagues worked hard to estimate and reduce uncertainties in the calculations. As he remarked only six years after the first solar-neutrino calculations, "Our best solar model, using what we consider the most likely set of parameters for the proton–proton chain, predicts a capture rate of only 6 SNU..." This was almost a factor of 10 smaller than the original calculations, but it was still about three times larger than the results from Davis's experiment. "We estimate that the uncertainty in the new capture rate is roughly a factor of two or three...," Bahcall remarked. "About 80% of the expected rate of 6 SNU represents neutrinos from the decay of radioactive boron, ^8B..." (Bahcall 1969).

On the experimental side, over the next several decades Davis and his colleagues upgraded and refined the experimental apparatus, with the continuing support of Brookhaven until Davis's retirement in 1984, and subsequently with support through the University of Pennsylvania, where Davis continued his work for another 18 years. These improvements ultimately enabled the experimenters to measure the flux of neutrinos from the Sun with a precision of 5% (Cleveland et al. 1998). For his work on this challenging experiment, Davis shared the Nobel Prize in 2002.

Despite many improvements in both theory and experiment, however, the measured solar neutrino rate stubbornly persisted in remaining maddeningly low, about one third of the best theoretical results. What could be wrong? This frustratingly unresolved question led to the development of several new types of solar neutrino detectors in the 1980s and 1990s.

The first, a Japanese experiment called Kamiokande, led by Masatoshi Koshiba (1926–2020) and Yoji Totsuka (1942–2008), employed a large volume of highly purified water as the detector medium. "The Kamiokande detector (http://en.wikipedia.org/wiki/Super-Kamiokande), begun in 1982 and completed a year later, was a 3000-ton water Čerenkov detector built to search for evidence of proton decay. It was upgraded in 1985 to enable it to detect solar neutrinos as well" (Bahcall 2004). This experiment measured "the rate at which electrons in the water scattered the highest-energy neutrinos emitted from the Sun. The water detector was very sensitive, but only to high-energy neutrinos that are produced by a rare nuclear reaction (involving the decay of the nucleus ^8B) in the solar energy production cycle. The original Davis experiment with chlorine was primarily, but not exclusively, sensitive to the same high-energy neutrinos" (Bahcall 2004).

In 1988, Kamiokande did succeed in detecting solar neutrinos and "confirmed that the number of neutrino events ... observed was less than predicted by the theoretical model of the Sun and by the textbook description of neutrinos..." (Bahcall 2004). Koshiba also shared the 2002 Nobel Prize for this work. The detector was upgraded in 1996, increasing the amount of water it contained by a factor of 15 and incorporating 10 times as many phototubes as the original. This much-more-sensitive detector was renamed "Super-Kamiokande." It continued to measure the solar neutrino flux into the 21st century (Fukuda et al. 1996). "Super-Kamiokande (led by Totsuka and Yochiro Suzuki), made more precise measurements of the higher energy neutrinos and confirmed the original deficit of higher energy neutrinos found by the chlorine and Kamiokande experiments" (Bahcall 2004).

Also in the 1980s, a collaboration of Canadian, American, and British physicists designed a new solar-neutrino detector, using 1000 tons of heavy water (D_2O) as the detector medium.[4] Construction began in 1990 at a site 6800 feet underground in a nickel mine near Sudbury, Ontario. Led by Canadian Arthur McDonald, it was named "SNO" (pronounced "snow"), standing for "Sudbury Neutrino Observatory." "The new detector ... was able to study in a different way the same higher-energy solar neutrinos that had been investigated previously in Japan with the Kamiokande and Super-Kamiokande ordinary-water detectors.... For their first measurements, the SNO collaboration used the heavy-water detector in a mode that is sensitive only to electron neutrinos. The SNO scientists observed approximately one-third as many electron neutrinos as the standard computer model of the Sun predicted to be created in the solar interior. The Super-Kamiokande detector, which is primarily sensitive to electron neutrinos but has some sensitivity to other neutrino types, observed about half as many events as were expected" (Bahcall 2004).

Two additional solar neutrino experiments, both employing gallium as the detector medium, also came into operation in the 1990s. The Soviet–American Gallium Experiment (SAGE), led by Till Kirsten in Germany, operated from 1990 to 2007 and employed as a target more than 50 tons of liquid gallium in a laboratory deep under the Caucasus Mountains in Russia. The observed flux of solar neutrinos was less than 60% of that predicted by standard solar models (http://en.wikipedia.org/wiki/SAGE_(Soviet%E2%80%93American_Gallium_Experiment)). The European GALLEX ("Gallium Experiment") project, led by Vladimir Gavrin in Russia, operated from 1991 to 1997 in the Gran Sasso underground laboratory in Italy. It employed the reaction $^{71}Ga + v_e \rightarrow {}^{71}Ge + e^-$, with a threshold energy of 0.23 MeV, to search for evidence of neutrinos from the p–p reaction in the solar core. During its six-year run, the experiment detected less than the number predicted by standard solar models (http://en.wikipedia.org/wiki/GALLEX). "The fact that GALLEX and SAGE were sensitive to lower energy neutrinos was very important since ... [theorists] ... could calculate more accurately the number of low energy neutrinos than the number

---

[4] Deuterium (D) is a heavy isotope of hydrogen, in which the nucleus of the hydrogen atom is a deuteron (d) rather than a proton (p).

of higher energy neutrinos.... So both high and low energy neutrinos were missing, although not in the same proportions" (Bahcall 2004).

After the initial SNO experiments, operating in a mode sensitive only to electron neutrinos, had confirmed the deficit of solar neutrinos found by earlier work, the SNO experimentalists conducted further investigations, operating the detector in a mode sensitive to all three neutrino types. "Combining the SNO and Super-Kamiokande measurements, the SNO collaboration determined the total number of solar neutrinos of all types (electron, muon, and tau) as well as the number of just electron neutrinos. The total number of neutrinos of all types agrees with the number predicted by the computer model of the Sun. Electron neutrinos constitute about a third of the total number of neutrinos" (Bahcall 2004). As Bahcall remarked with evident gratification, "This shows that we understand how the Sun shines, the original question that initiated the field of solar neutrino research" (Bahcall 2004).

These "epochal results announced in June 2001 (Ahmad et al. 2001) were confirmed by subsequent measurements.... [The] results from the SNO measurements alone show that most of the neutrinos produced in the interior of the Sun, all of which are electron neutrinos when they are produced, are changed into muon and tau neutrinos by the time they reach Earth" (Bahcall 2004).

The solution to the solar neutrino mystery thus proved to reveal *not* a problem with the modern theoretical models for the Sun but instead a deficiency in our understanding of the fundamental properties of neutrinos. This had actually been anticipated in a prophetic paper by Italian-born physicist Bruno Pontecorvo (1913–1993) in 1967 (Pontecorvo 1967). There, he "first discussed the possibility of solar neutrino oscillations... [but] dismissed the possibility of using solar neutrinos to test for neutrino oscillations, because—he believed—that the theoretical expectations were not reliable to the required accuracy" (Bahcall 2004). A subsequent visionary paper by Vladimir N. Gribov (1930–1997) and Pontecorvo (Gribov and Pontecorvo 1969) "proposed the basic idea underlying the correct solution of the solar neutrino problem. More than three decades were needed in order to prove that indeed new particle physics was required to explain what happened to the uncounted neutrinos" (Bahcall 2004). "Lower energy solar neutrinos switch from electron neutrino to another type as they travel in the vacuum from the Sun to the Earth. The process can go back and forth between different types. The number of ... oscillations ... depends upon the neutrino energy. At higher energies, the process of oscillation is enhanced by interactions with electrons in the Sun or in the Earth. Stas Mikheyev, Alexei Smirnov, and Lincoln Wolfenstein first proposed (Wolfenstein 1978; Mikheyev & Smirnov 1985) that interactions with electrons in the Sun could ... cause the neutrinos to oscillate more vigorously between different types.... The solution of the mystery of the missing solar neutrinos is that neutrinos are not, in fact, missing. The previously uncounted neutrinos are changed from electron into muon and tau neutrinos that are more difficult to detect.... Solar neutrinos ... are created as electron neutrinos in the Sun, but on the way to the Earth they change their type... a quantum mechanical process, called 'neutrino oscillations'" (Bahcall 2004).

Bahcall has provided a fitting epilogue to this story. He said,

I am astonished when I look back on what has been accomplished in the field of solar neutrino research over the past four decades. Working together, an international community of thousands of physicists, chemists, astronomers, and engineers has shown that counting radioactive atoms in a swimming pool full of cleaning fluid in a deep mine on Earth can tell us important things about the center of the Sun and about the properties of exotic fundamental particles called neutrinos. If I had not lived through the solar neutrino saga, I would not have believed it was possible (Bahcall 2004).

Even though the resolution of the solar-neutrino problem proved to be due to the physical properties of neutrinos rather than to an inadequate model for the Sun, the decades-long effort to identify and eliminate uncertainties in the physics of the solar interior produced substantial improvements in solar models. Recall that the 1963 predictions of solar-neutrino fluxes, based on the best theoretical models of the time, yielded 50 SNU, more than 10 times larger than the experimental results from the Davis experiment. In contrast, by 2005, the theoretically calculated flux of neutrinos from the p + p reaction in the center of the Sun agreed perfectly with the experimentally determined value, and the uncertainties in the results were reduced to 2% for the experiments and 1% for the theoretical model calculations (Bahcall et al. 2005).

Figure 8.4 shows how determinations of the central temperature of the Sun have been improved by the efforts to reconcile theory with the solar-neutrino experiments.

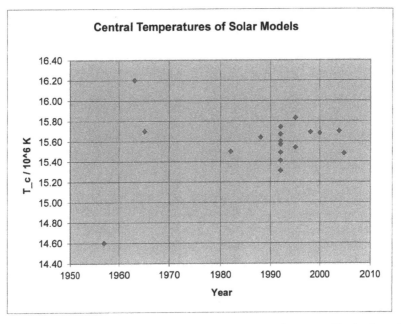

Figure 8.4. Central temperatures of various solar models from 1957 to the present. Plot by the author from references cited in the text.

The earliest point in the figure represents the central temperature of a model computed in 1957 (Weyman 1957), well before any of the solar-neutrino experiments had been carried out. The second point is the central temperature of the model used by Bahcall et al. for their original 1963 calculation of the solar-neutrino flux. The remaining points represent models computed by different groups using differing assumptions about various physical properties of solar matter (Strömgren 1965). The average since 1982 is 15.59 ± 0.13 million degrees Kelvin; in other words, the central temperature of the Sun is now known to an accuracy slightly better than 1% as a result of the solar-neutrino measurements.

During the same decades in which the solar-neutrino problem was being investigated so intensively, a completely different kind of observational test of models for the solar interior also emerged. As described in the following chapter, it proved to have the potential to constrain other aspects of the structure of the Sun. Combining the results from the solar-neutrino experiments with those from helio-seismology—as the new approach came to be called—resulted in a substantial improvement in our knowledge of the interior of our nearest star, as discussed in more detail at the end of Chapter 9.

References

Ahmad, Q. R., et al. 2001, PhRvL, 87, 1301

Bahcall, J. N. 1964, PhRvL, 12, 300

Bahcall, J. N. 1969, SciAm, 221, 28

Bahcall, J. N. 2004, http://www/nobelprize.org/nobel_prizes/themes/physics/bahcall/

Bahcall, J. N., Fowler, W. A., Iben, I. Jr., & Sears, R. L. 1963, ApJ, 137, 344

Bahcall, J. N., Pinsonneault, M. H., & Basu, S. 2001, ApJ, 555, 990

Bahcall, J. N., Serenelli, A. M., & Basu, S. 2005, ApJ, 621, L85

Cleveland, B. T., et al. 1998, ApJ, 496, 505

Cowan, C. L. Jr., Reines, F., Harrison, F. B., Kruse, H. W., & McGuire, A. D. 1956, Sci, 124, 103; see also

    Reines, F. 1995, "The Neutrino: From Poltergeist to Particle," Nobel Lecture

Davis, R. Jr. 1964, PhRvL, 12, 303

Davis, R. Jr., Harmer, D. S., & Hoffman, K. C. 1968, PhRvL, 20, 1205

Fermi, E. 1934, ZPhy, 88, 161

Fukuda, Y., et al. 1996, PhRvL, 77, 1683; see also

    Fukuda, S., et al. 2001, PhRvL, 86, 5651

Gribov, V. N., & Pontecorvo, B. M. 1969, PhLB, 28, 493

http://en.wikipedia.org/wiki/GALLEX; accessed 14 April 2015

http://en.wikipedia.org/wiki/John_N_Bahcall; accessed 12 March 2015

http://en.wikipedia.org/wiki/Raymond_Davis_Jr; accessed 12 March 2015

http://en.wikipedia.org/wiki/SAGE_(Soviet%E2%80%93American_Gallium_Experiment); accessed 14 April 2015

http://en.wikipedia.org/wiki/Super-Kamiokande; accessed 14 April 2015

Lande, K. 2006, PhT, 59, #10, 78

Mikheyev, S. P., & Smirnov, A. Y. 1985, SvJNP, 42, 913

Overbye, Dennis 19 August 2005, *The New York Times*, "John N. Bahcall, 70, Dies; Astrophysicist at Princeton."

Pauli, W. 1934, Rapports du Septième Conseil du Physique Solvay, Brussels, 1933 (Paris: Gautier-Villars et Cie)

Pontecorvo, B. M. 1967, ZETF, 53, 1717

Preston, M. A. 1962, Physics of the Nucleus (Reading, MA: Addison-Wesley) 389

Strömgren, B. 1965, Stars and Stellar Systems. VIII. Stellar Structure, ed. L. H. Aller, & D. B. McLaughlin (Chicago, IL: Univ. Chicago Press) 269; see also

Guenther, D. B., et al. 1992, ApJ, 387, 372

Bahcall, J. N. 2015, models computed from 1982 to 2005, collected at http://www.sns.ias.edu/~jnb/; accessed 4 July 2015

Weyman, R. 1957, ApJ, 126, 208

Wolfenstein, L. 1978, PhRvD, 17, 2369

Chapter 9

The Sounds of Sunlight

In 1960, Caltech physicist Robert B. Leighton (1919–1997) took a series of photographic measurements of Doppler velocities of the gases at the surface of the Sun. He expected to find evidence of motions associated with the solar granulation—the random boiling motions at the top of the solar convection zone. Instead, he discovered (Leighton 1960) vertical motions with a strong oscillatory character and a period of (296 ± 3) seconds, or (4.93 ± 0.05) minutes. Not surprisingly, these motions became known as the solar "five-minute oscillations." And in 1972, Charles L. Wolff suggested that the 5-minute oscillations are actually nonradial pulsations of the entire Sun (Wolff 1972).

Several decades earlier, British astrophysicist Thomas G. Cowling (1906–1990) had shown that a star can support two main classes of global oscillation modes (Cowling 1941). The so-called "p-modes"—for which the pressure of stellar matter provides the restoring force for the pulsations—are essentially acoustic oscillations —sound waves—that travel through the body of the star. Just as a sound wave propagating in an organ pipe on Earth produces a particular tone (frequency) when the wavelength fits the length of the pipe, so too an acoustic wave in a star produces a resonance at a particular frequency when the wave fits exactly into the interior of the star. And just as waves with increasing numbers of nodes inside an organ pipe produce overtones of higher frequencies, so too do acoustic waves with increasing numbers of nodes inside a star produce resonances at higher-overtone frequencies.

The second class of modes Cowling termed "g-modes." The restoring force for this type of mode is provided by thermally induced buoyancy caused by the oscillation. While the periods of p-modes get shorter (the frequencies get higher) as the number of nodes (zero-crossings) increases, the periods of the g-modes get longer as the number of nodes increases. For solar models, p-modes all have periods shorter than about 64 minutes, while g-modes all have longer periods (Christensen-Dalsgaard 1982).

Mathematically, a stellar oscillation mode can be represented as a function that depends upon the radial distance from the center of the star multiplied by a function of latitude and longitude. The radial function has some number of nodes (radial locations where the function goes through zero), represented by an integer n. A mode with $n = 0$ has no nodes in the radial function; one with $n = 1$ has a single node somewhere between the center and the surface; and so on. The higher the value of n, the more "wiggly" the function is. Similarly, the functions of latitude and longitude —called "spherical harmonics"—have nodes represented by the integers l and m. A spherical harmonic with $l = 0$ and $m = 0$ has no nodes. One with $l = 1$ has a single node (the "equator") separating the stellar surface into two hemispheres ("north" and "south"); it is called a "dipole" mode. A mode with $l = 2$ is a "quadrupole" mode. Non-zero values of m correspond to nodes in longitude. Figure 9.1 illustrates the structure of a solar oscillation mode characterized by a spherical harmonic with a specific value of l and m.

Figure 9.1. Artist's conception of a solar oscillation mode. The right-hand side of the image shows a cutaway view of the solar interior, with the outer stippled region representing the subsurface convection zone, and the colors shading from orange to yellow representing temperature increase toward the center through the region dominated by the flow of radiation. The red and blue areas represent regions with slightly higher or lower pressure (for the p-modes) or with slightly higher or lower temperature (for the g-modes). Because each mode is a strictly periodic oscillation, the red regions become blue on the next half cycle, and vice versa. Credit: GONG/NSO/AURA/NSF; source: gong.nso.edu/info/helioseismology.html.

In an effort to clarify the nature of the solar 5-minute oscillations, in September 1974 Franz-Ludwig Deubner (Deubner 1975) undertook a series of new photo-electric observations of the Sun using a solar telescope in Anacapri, an island off the coast of Naples, Italy. He measured the Doppler velocities of light from the central part of the solar disk, which he scanned periodically for several hours. He allowed for the effects of solar rotation, so that his observations remained focused on the same elements of the solar surface. Deubner then analyzed the data using the Fast Fourier Transform (FFT) algorithm to obtain the characteristic frequencies ω and horizontal wavenumbers k associated with the Doppler velocity measurements.[1]

Plotting the results as contours showing the distribution of signal strength in a so-called "diagnostic diagram" (ω vs. k), Deubner found a "number of more or less parallel ridges running in a roughly diagonal direction at regularly spaced intervals..." Comparing his results with theoretical predictions, Deubner concluded that his observations "have shown that the 5-minute oscillations may in fact be interpreted as low wavenumber nonradial acoustic eigenmodes of the subphoto-spheric layers of the solar atmosphere." This indicates, he continued, "that substantial parts of the solar interior participate in the oscillation. A direct comparison of our observations with the spectra of free ... normal modes of oscillation of the whole Sun has not yet been possible because of lack of spectral resolution [in the observational data]. But the necessary resolution can certainly be achieved in the near future, putting at our disposal an excellent diagnostic tool which allows us to probe the internal structure of our Sun..."

Before they could make use of the 5-minute oscillations to probe the interior of the Sun, however, solar researchers had to make substantial advances in the amount and quality of observational data, in theoretical models of the solar interior, and in methods of analysis.

To obtain enough data of sufficient quality, solar observers had to obtain long records of uninterrupted data. However, at any given observatory site, the daily recurrence of nighttime hours—during which the Sun of course cannot be observed—posed a problem. These regularly recurring gaps in the data masked some of the periodicities the scientists were trying to detect and study. One solution to this problem was to carry out observations from the Earth's South Pole during the austral summer, because the Sun never sets there during these months. This approach enabled observers to extend the (nearly) continuous data records over more than a month. Another approach was to establish a small number of identical observing stations spaced around the world in longitude, so that as the Sun set at one location, the observations could be "handed off" to another observing station farther west. This approach in principle allows data records to be created that may reach months or years in length. Yet another avenue was to conduct observations from spacecraft dedicated to solar research. In the end, all three approaches have been employed, and each has contributed to the creation of extensive data sets of

---

[1] Low frequency—small ω—corresponds to long oscillation periods, and low wavenumber k corresponds to long wavelengths.

extremely high quality, which in turn have provided the foundation for detailed seismic investigations of the Sun.

The first observations of solar oscillations from the South Pole were whole-disk measurements reported in 1980 (Grec et al. 1980). The advent of CCD and CID detectors[2]—which enabled the recording of images of the entire solar disk with high spatial resolution through suitable Doppler-velocity filters in rapid time sequence—transformed the observational study of solar oscillations. T. L. ("Tom") Duvall, Jr., John W. ("Jack") Harvey, Martin Pomerantz (1916–2008), and their collaborators used these detectors to continue the "helioseismic" observations—as they had by then come to be called—from the South Pole. During the austral summer of 1981–82, for example, they obtained 160 hours of data every 90 seconds. These types of observations continued until about 1995, when high-quality data from ground-based helioseismic networks and spacecraft observations became available.

The use of a network of solar observing stations distributed around the world was pioneered by the solar research group at the University of Birmingham in England. Beginning in the mid-1970s, this group began whole-disk measurements of the solar p-mode oscillations. Initially the observations were carried out at Izana on Tenerife in the Canary Islands. When it became clear that 24-hour coverage was essential to provide sufficiently accurate measurements of the spectrum, the group established a second observing station on Haleakala in the Hawaiian Islands in 1981. Over the following decade, they installed additional automatic stations, ultimately employing a network of six observing stations, named the Birmingham Solar Oscillation Network (BiSON). Because the BiSON instruments perform whole-disk measurements, they are only sensitive to modes with $l < 3$.

The need for data at higher l values led to discussions within the international helioseismic community during the early 1980s. At a workshop in 1984, these discussions crystallized into a specific proposal for a network of six automated observing stations spaced around the globe. Called the "Global Oscillation Network Group (GONG)," this project initially deployed a 256×242 pixel camera (later upgraded to 1024×1024), which obtain solar velocity images every 20 minutes. The GONG project is a truly international effort, with 175 scientists coming from 70 institutions in more than 20 countries (Leibacher 1999). The six GONG sites (see Figure 9.2) came online in 1995 and are continuing to provide data into the 21st century with an average duty cycle over the network of almost 90%.

Also in 1995, the Solar and Heliospheric Observatory (SOHO) was launched into space. Since 1996, it has been "on station" near the inner Lagrange point, L1,[3] located between the Earth and the Sun. SOHO carries three instruments dedicated to helioseismic observations: the Global Oscillations at Low Frequencies (GOLF) instrument, the Michelson Doppler Imager (MDI), and the Variability of solar Irradiance and Gravity Oscillations (VIRGO) instrument. The GOLF, VIRGO, and

---

[2] A CCD is a "charge-coupled device," while a CID is a "charge-injection device." The former erases the data in each picture element ("pixel") as it is read out, while the latter does not.

[3] The point at which the gravitational forces from the Earth and the Sun balance in a rotating coordinate frame centered on the Sun and with a period of one year.

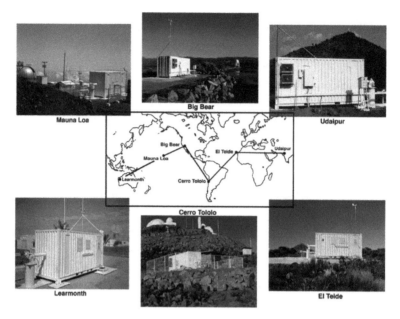

Figure 9.2. The six GONG sites. The map in the center of the figure shows the geographic locations of each station, and the six surrounding images show their appearances. Credit: GONG/NSO/AURA/NSF; source: gong.nso.edu/gallery/disk2k10/data/resource/sites/sitemap.jpg.

ground-based BiSON facilities are designed to measure oscillations with low values of the spherical harmonic index l, which probe deep into the solar core, while the MDI and ground-based GONG instruments cover l values up to about 150. The result of these observations has been the detection of a well-resolved diagnostic diagram (see Figure 9.3), now plotting the p-mode oscillation frequencies ω against the spherical harmonic index l, rather than the horizontal wavenumber k, to facilitate comparison with frequencies calculated from theoretical models.

As solar observers began to identify more and more oscillation modes, with increasingly accurate frequency determinations, pulsation theory experts like Douglas Gough (Cambridge University, England), Jørgen Christensen–Dalsgaard (Aarhus Universitet, Denmark), Wojciech Dziembowski (Copernicus Astronomical Center, Poland), and their collaborators began to develop methods to utilize these observational data to infer the distributions of physical properties inside the Sun. They adapted some techniques that had been pioneered by geophysicists to infer the distributions of density and other physical quantities inside the Earth, and they developed other methods based on stellar pulsation theory. Gough provided a lucid explanation of these inversion methods at a conference in Boulder, Colorado, in 1982 (Gough 1982) and in a subsequent technical paper (Gough 1985).

While helioseismic inversion methods were being developed and tested, other scientists worked to improve the solar models. Some created new models for the thermodynamic properties of stellar matter—the so-called "equation of state"—especially for the challenging regions in which atoms are only partially ionized.

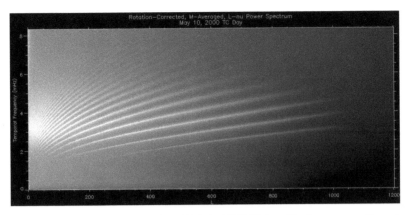

Figure 9.3. Diagnostic ω–l diagram from 1024 × 1024 pixel GONG data. The frequency ω of a mode, which is the inverse of its period, is plotted on the vertical axis; a frequency of 3 millihertz (mHz) corresponds to a period of about 5.5 minutes. The horizontal axis represents the angular degree l of the spherical harmonic. The areas of the diagram are brighter for the strongest motions and darker for the weakest. Each band in the figure corresponds to a single value of the radial index n, with the value of n increasing from the bottom to the top of the figure. Because the frequencies increase as n increases, these oscillations are solar p-modes. Credit: GONG/NSO/AURA/NSF; source: gong.nso.edu/gallery/disk2k10/data/resource/spectra/lnu2.jpg/.

Others developed new opacities for stellar matter or improved values for thermonuclear reaction rates. Still others explored the effects produced by the gradual diffusive settling of helium and heavier elements toward the center of the Sun under the influence of its strong gravity. And some re-determined the abundances of the chemical elements in the Sun.

As new determinations for the various physical properties of solar matter became available, researchers at Aarhus Universitet in Denmark, Princeton and Yale Universities in the United States, and others incorporated them into increasingly refined theoretical models for the solar interior. By the latter part of the 1990s, when high-quality data from ground-based helioseismology networks began to become available, solar models had been improved considerably. Comparisons between helioseismic inversions and data from the models then provided exacting tests of the theoretical calculations. A few of these are summarized below.

9.1 Distribution of Sound Speed in the Sun

In 1985, Jørgen Christensen-Dalsgaard and his collaborators demonstrated a method for determining the distribution of the sound speed c in the solar interior by inverting the observed p-mode frequency distributions (Christensen-Dalsgaard et al. 1985). Because the square of the sound speed, c^2, is proportional to the local temperature, these measurements yield critical insights about the temperature distribution inside the Sun. Figure 9.4 shows the relative difference in c^2 as determined in the late 1990s by inverting four months of GONG data for the frequencies of p-modes with values of l ranging from 0 to 150 (Anderson et al. 1997) with the same quantity obtained from a standard solar model (Christensen-Dalsgaard et al. 1996). The plotted quantity—labeled "δc^2"—is the square of the

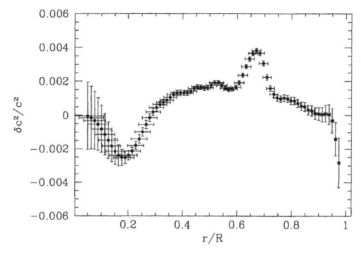

Figure 9.4. Relative difference in the square of the sound speed in the solar interior as determined from helioseismic inversion of several months of GONG data compared to a solar model. The horizontal and vertical bars show the error estimates for each point. Because the square of the sound speed is proportional to the local temperature, this figure shows that that temperature distributions in the best current solar models are better than 0.5%. Source: Anderson, E., et al. 1997, in Provost and Schmider 1997, p. 151; used under license number RP2023063SM from Springer Nature.

sound speed in the Sun, as determined from the helioseismic inversions, minus the square of the sound speed in the solar model. If the model perfectly reproduced the measured values of c^2, this plot would be a horizontal line at $\delta c^2/c^2 = 0$. The non-zero values of this quantity demonstrate that the best theoretical models of the time were not perfect. (Nor are they today.) However, it is important to recognize that the measured deviations are very small; the differences between the actual values of c^2 in the Sun and those in the model are everywhere less than 0.4%. Thus, although not perfect, the theoretical models are actually in exceptionally good agreement with the observational determinations. Our understanding of the solar interior thus rests on a very firm foundation of established physical theory.

The details of the figure also reveal three interesting features, which provide clues that may enable further improvements in the models. First, there is a prominent peak in $\delta c^2/c^2$ just below the base of the convection zone at about 0.7 R_{Sun} (see below). Second, in the radiative region between about 0.3 R_{Sun} and 0.7 R_{Sun}, the actual sound speed in the Sun is slightly greater than it is in the model. And third, below this radius—in the region that generates 90% of the Sun's luminosity—the sound speed is slightly lower than it is in the model.

9.2 The Depth of the Solar Convection Zone

At the base of the solar convection zone, the temperature gradient changes abruptly —discontinuously, in theoretical models—from the nearly adiabatic gradient that is associated with efficient convection to the radiative gradient that prevails through-out the rest of the solar interior. In 1991, Christensen–Dalsgaard and his colleagues detected this change in temperature gradient in the helioseismological data and showed that it can be used to establish the depth of the solar convection zone rather precisely (Christensen–Dalsgard et al. 1991). They found that the inner boundary of the convection zone lies at 0.713 ± 0.003 solar radii. This measurement also confirmed that the opacity just beneath the convection zone is about 10% larger than earlier calculations at Los Alamos had indicated, but it is consistent with new opacity calculations (Iglesias & Rogers 1991) at Lawrence Livermore National Laboratory. Interestingly, the new Livermore opacities not only improved the standard solar model but also helped resolve discrepancies in other areas of astrophysics.

9.3 The Density Distribution in the Sun

Solar researchers also compared the relative difference, $\delta\rho/\rho$, between the local mass density in the Sun, as determined from helioseismic inversions of GONG data, with that in a standard solar reference model (see Figure 9.5). The relative error in the density is everywhere less than 2% in magnitude. As the figure shows, the actual density is slightly lower at the center of the Sun than in the model, and it is slightly higher in the outer layers, indicating that the Sun is slightly less centrally condensed than the model indicates. A dozen years later, Sarbani Basu and his collaborators published determinations of the internal density and sound-speed distributions in the Sun using data from BiSON and from the Michelson Doppler Imager (MDI) instrument aboard the SOHO spacecraft (Basu et al. 2009), and they compared them with results from newer model calculations. They found the relative error in $\delta c^2/c^2$ to be less than 0.5% everywhere, *except* just below the base of the convection zone,

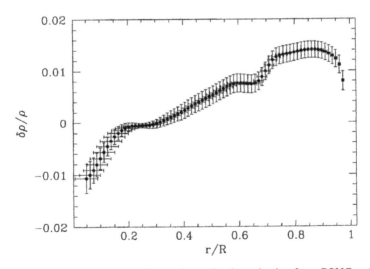

Figure 9.5. Relative difference between a helioseismic density determination from GONG and the density distribution in a standard solar model. This figure shows that the model densities agree with the helioseismic results better than 2% everywhere except in the subsurface convection zone. Source: Anderson, E., et al. 1997, in Provost and Schmider 1997, p. 151; used under license number RP2023063SM from Springer Nature.

where a prominent peak like that in Figure 9.4 remains. Their comparison between the densities determined from several different helioseismic inversions with those from a newer standard solar model (model BP04; Bahcall & Pinsonneault 2004) are generally similar to those shown in Figure 9.5. In all cases, the actual density in the solar convection zone is larger than in the model, as indicated in Figure 9.5. The upshot, again, is that the remaining errors in the solar models are of the order of a few percent or less.

9.4 Distribution of Angular Velocity in the Sun and the Solar Dynamo

In the four centuries since Galileo first discovered the existence of dark spots on the solar surface and utilized daily observations of their motions across the face of the Sun to determine its rotation rate, the rotation of the solar surface has been well-established. In particular, these observations showed that the Sun does not rotate as a solid body. At the equator, the rotation period is 25.8 days, while it increases to about 28.0 days at latitude 40° and 36.4 days at 80° (Abell et al. 1987). Aside from theoretical inferences, however, nothing was known directly about the distribution of angular velocities within the Sun until the advent of helioseismology.

This changed dramatically in 1984, when Tom Duvall, Jack Harvey, and their collaborators used the observed differences between prograde and retrograde[4] solar oscillations to determine the distribution of rotation rates[5] Ω inside the Sun. They concluded (Duvall & Harvey 1984) that "most of the Sun's volume rotates at a rate close to that of the surface, but also that the energy-generating core may rotate more rapidly than the surface." About the same time, Douglas Gough developed a different method to determine the internal solar rotation near the Sun's equatorial plane (Gough 1984) and found that it "yields a result similar to that derived by Duvall et al. ... by another procedure."

Five years later, Tim Brown and his colleagues (Brown et al. 1989) utilized the solar oscillation data that he and Cherilynn Morrow had obtained in 1987 at the National Solar Observatory's facility in Sunspot, New Mexico, to investigate the internal angular velocity of the Sun. They used several different inversion methods and found that "All of the results are in substantial agreement: a latitudinal variation of Ω similar to that observed at the solar surface exists throughout the Sun's convection zone... [while] in the radiative interior, the equatorial and polar rotation rates approach a common value, intermediate between their surface values..."

A decade after that, as long time-series data from GONG and the LOWL instrument in Hawaii became available, researchers were able to refine the angular velocity determinations throughout the solar interior. They found that (Charbonneau et al. 1999) "there exists very little *radial* shear in the solar convective

---

[4] A "prograde" oscillation mode propagates around the surface of the Sun in the same direction as the solar rotation, while a "retrograde" mode travels in the opposite direction.
[5] The rotation rate Ω equals 2π divided by the rotation period.

envelope... Instead, differential rotation in the bulk of the envelope is primarily latitudinal..., with the angular velocity at a given latitude approximately equal to its surface value. It also became apparent that ... the underlying radiative core ... is in a state of near solid-body rotation ... The transition to the overlying differentially rotating envelope occurs across a thin ($\leqslant 0.1\ R_{Sun}$), approximately spherical layer often now called the *solar tachocline*..." Further, "Because of the strong radial shear present in its equatorial region, as revealed by helioseismology, the solar tachocline is also a promising location for the seat of the solar dynamo."

These results were a surprise to theorists. "Unlike the predictions of stellar-evolution models, the radiative interior [of the Sun] is found to rotate roughly uniformly. The rotation within the convection zone is also very different from prior expectations, which had been that the rotation rate would depend primarily on the distance from the rotation axis. Layers of rotational shear have been discovered at the base of the convection zone ..." (Thompson et al. 2003). The observational results were emphatically confirmed in 2008, by detailed analyses of 2088-day time series from the MDI and GOLF instruments aboard the SOHO spacecraft combined with ground-based data from GONG. They showed (Eff–Darwich et al. 2008) "the Sun rotating like a rigid solid in most of the radiative zone, and slowing down in the core ($r/R_{Sun} < 0.2$)." They also refined the latitudinal variation of the rotation rate through the convection zone (see Figure 9.6).

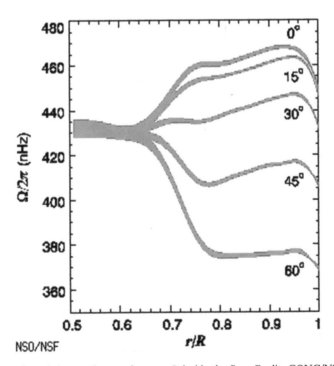

Figure 9.6. Distribution of the angular rotation rate Ω inside the Sun. Credit: GONG/NSO/AURA/NSF; source: gong.nso.edu/gallery/disk2k10/data/resource/torsional/torsional.html.

Several of these results are especially noteworthy. First, as noted above, observational determinations of the internal rotation rates differ markedly from prior theoretical expectations. This serves as a cautionary warning: theoretical predictions must always be tested against observation or experiment. Thanks to the detailed observational measurements, we now have reliable knowledge of the actual distribution of rotation within the Sun. Second, because few of the oscillation modes that can be observed at the solar surface penetrate deep into the core, the rotation rate there still remains uncertain. Early determinations seemed to indicate that the core rotates somewhat more rapidly than the bulk of the solar interior, while the most recent determinations seem to indicate that it may be rotating somewhat more slowly. Third, the location of a strong shear layer in the solar tachocline, which may therefore be the seat of the solar dynamo, approximately coincides with the location of the peak in $\delta c^2/c^2$. This may indicate that the two are connected, but a more detailed analysis of the solar dynamo is likely to be needed to determine this. Because the solar rotation rate is slow enough that it does not produce large deviations from spherical symmetry, however, our basic understanding of the internal structure of the Sun—and by extension of other stars as well—remains on a solid foundation.

The resolution of the solar neutrino problem and the inversion of solar oscillation data to determine some of the physical quantities inside the Sun have demonstrated that we really do know the internal properties of the very nearest star to an accuracy of a few percent or better. Scientists have even been able to derive a helioseismic age for the Sun of (4.66 ± 0.11) billion years (Dziembowski et al. 1999)—improved a few years later to (4.57 ± 0.11) billion years (Bonanno et al. 2002)—entirely consistent with the age of (4.57 ± 0.02) billion years obtained from meteorites.[6] To be sure, our knowledge is not perfect, but the current level of accuracy in our understanding of the properties of matter and the structure of the Sun—tested and refined by the observational data—provides confidence in using the same theoretical foundation to explore the properties of other Main Sequence stars, and we turn to that in the following chapter.

References

Abell, G. O., Morrison, D., & Wolff, S. C. 1987, Exploration of the Universe (Philadelphia, PA: Saunders College Publishing) 497

Anderson, E., et al. 1997, in Provost and Schmider 1997 151

Bahcall, J. N., & Pinsonneault, M. H. 2004, PhRvL, 92, 121301; see also http://www.sns.ias.edu/~jnb/SNdata/Export/BP2004/bp2004stdmodel.dat

Basu, S., Chaplin, W. J., Elsworth, Y., New, R., & Serenelli, A. M. 2009, ApJ, 699, 1403

Bonanno, A., Schlattl, H., & Paternò, L. 2002, A&A, 390, 1115

Brown, T. M., Christensen–Dalsgaard, J., Dziembowski, W. A., Goode, P., Gough, D. O., & Morrow, C. A. 1989, ApJ, 343, 526

Charbonneau, P., et al. 1999, ApJ, 527, 445

---

[6] The meteoritic age of the Sun is (4.57 ± 0.02) billion years, according to Bahcall, J. N., Pinsonneault, M. H., and Wasserburg, G. J. 1995, RvMP, **67**, 781.

Christensen–Dalsgaard, J. 1982, MNRAS, 199, 735

Christensen–Dalsgaard, J., et al. 1985, Natur, 315, 378

Christensen–Dalsgaard, J., et al. 1996, Sci, 272, 1286

Christensen–Dalsgard, J., Gough, D. O., & Thompson, M. J. 1991, ApJ, 378, 413; see also
    Guzik, J. A., & Cox, A. N. 1993, ApJ, 411, 394

Cowling, T. G. 1941, MNRAS, 101, 367

Deubner, F.-L. 1975, A&A, 44, 371

Duvall, T. L. Jr., & Harvey, J. W. 1984, Natur, 310, 19; see also
    Duvall, T. L. Jr., et al. 1984, Natur, 310, 22
    Howe, R. 2009, LRSP, 6, 1

Dziembowski, W. A., Firoentini, G., Ricci, B., & Sienkiewicz, R. 1999, A&A, 343, 990

Eff–Darwich, A., Korzennik, S. G., Jiménez–Reyes, S. J., & García, R. A. 2008, ApJ, 679, 1636

Gough, D. O. 1982, Pulsations in Classical and Cataclysmic Variable Stars, ed. J. P. Cox, & C. J.
    Hansen (Boulder, CO: JILA and NBS) 117

Gough, D. O. 1984, RSPTA, 313, 27

Gough, D. 1985, SoPh, 100, 65

Grec, G., Fossat, E., & Pomerantz, M. 1980, Natur, 288, 541

Iglesias, C. A., & Rogers, F. J. 1991, ApJ, 371, 408

Leibacher, J. W. 1999, AdSpR, 24, 173

Leighton, R. B. 1960, in Thomas 1961 321; see also
    Noyes, R. W., & Leighton, R. B. 1963, ApJ, 138, 631

Thompson, M. J., Christensen–Dalsgaard, J., Miesch, M., & Toomre, J. 2003, ARA&A, 41, 599

Wolff, C. L. 1972, ApJ, 177, L87

Part III

From the Sun to the Stars

The same physical concepts that have enabled us to understand the interior of the Sun also apply to the stars. In this section, we find that there is both an upper limit and a lower limit to the objects that become Main Sequence stars. We also find that the more massive Main Sequence stars have such short H-burning lifetimes that they cannot have moved far from the places where they were formed. Studying these stellar nurseries has enabled astronomers to investigate how stars are actually formed. They have also found that objects can be formed with masses below the lower limit where H-burning reactions can occur, and we discuss these objects as well.

Chapter 10

Properties of Main Sequence Stars

From the studies of the Sun described in previous chapters, scientists have learned several important properties of Main Sequence stars: (1) They are essentially in hydrostatic equilibrium. That is, their physical properties—like the internal distributions of temperature and density—do not vary in time, except for small-amplitude disturbances like convection or pulsation. (2) They are mostly composed of hydrogen and helium, with a smattering of heavier elements like carbon, nitrogen, oxygen, iron, etc. (3) Thermonuclear hydrogen-burning—via either the proton–proton chain or the CNO cycle—provides the source of energy that sustains the stellar luminosity. (4) Energy is transported from the central regions where it is produced outward to the stellar surface either by the flow of radiation or by convection, and these different energy-transport processes dominate in different regions of the stellar interior. Let us now see what we can learn by applying these same physical principles to gaseous objects with masses either smaller or larger than the mass of the Sun.

The simplest reasonably realistic models for Main Sequence stars are for those that have just begun to burn hydrogen. Termed "zero-age Main Sequence"—or "ZAMS"—stars, these objects can be modeled using the same physical principles that have proven to work so successfully for the Sun. In their book on stellar structure (Hansen & Kawaler 1994), Carl Hansen (1933–2011) and Steve Kawaler tabulated a number of properties of ZAMS models. Some of the characteristics of these models are listed in Appendix C, and we use them as the basis for the discussion below. Stars with masses less than the solar mass are termed "lower Main Sequence" objects, while more massive stars are said to lie on the "upper Main Sequence."

10.1 The Lower Main Sequence

Stars on the lower Main Sequence are among the most common stars in the Milky Way Galaxy. Among the 100 nearest stars (Cox 2000), seven are of spectral class G, similar to that of the Sun, with effective temperatures ranging from about 5900 K

doi:10.1088/2514-3433/acce33ch10

down to about 5200 K and masses lying between about 1.0 and 0.8 solar masses. Another 16 are of the cooler spectral type K, with effective temperatures from about 5200 K down to about 3900 K and masses lying between about 0.8 and 0.5 solar masses. And a whopping 61 are of spectral type M, with temperatures less than about 3800 K and masses less than about 0.5 M_{Sun}. The nearest star, Proxima Centauri, lying at a distance of only 1.3 parsecs[1] and having a mass of about 0.1 M_{Sun}, belongs to this spectral class. In terms of sheer numbers, the lower Main Sequence dominates the stellar census hands down.

One difference from the solar case is that the dominant hydrogen-burning thermonuclear reactions in lower Main Sequence stars are those of the proton–proton chain; the carbon-cycle reactions identified by Bethe contribute negligibly because of the lower central temperatures of these low-mass stars; see Figure 10.1. The calculations for this figure assume that the relative mass fraction of heavy elements—including the carbon and nitrogen nuclei that catalyze the CNO bi-cycle reactions—is $Z = 0.0122$. For comparison with this figure, the central temperature of the Sun today is 15.48 million degrees Kelvin (or log $T = 7.19$ in terms of the variable used as the horizontal axis in this figure), just below the range of temperatures for which the CNO bi-cycle dominates the energy production in Main Sequence stars.

Another aspect of lower Main Sequence stars is that their surface convection zones grow progressively deeper as the stellar mass decreases (see Figure 10.2). This occurs because temperatures throughout a smaller-mass star are lower than they are in higher-mass stars. At these lower temperatures, the peak opacity increases, and

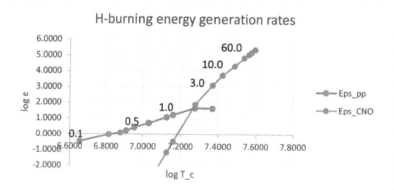

Figure 10.1. Variation with temperature of thermonuclear energy production by the proton–proton chain and the CNO bi-cycle along the ZAMS. The orange curve shows the p–p reaction rate, ε_{pp}, using the approximation from Hansen and Kawaler (1994), p. 238, and the gray curve shows the CNO reaction rate, ε_{CNO}, using their approximation on p. 241. The units of the energy-generation rates are ergs g^{-1} s^{-1}. The labels attached to individual points on the curves are the values in solar masses of the models with those central temperatures.

---

[1] About 270,000 times farther away than the Sun is from the Earth.

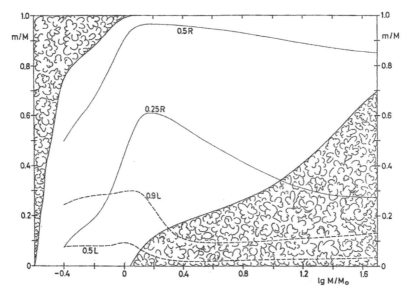

Figure 10.2. Convection zones in Main Sequence stars. The vertical axis gives the fractional mass m/M interior to that spherical shell in a star of given total mass M; it ranges from $m/M = 0$ at the center to $m/M = 1$ at the stellar surface. The horizontal axis is the logarithm of the total mass of a star in units of the solar mass. The value log $M/M_{Sun} = 0$ corresponds to a star with exactly the solar mass, while one with log $M/M_{Sun} = 1.0$ corresponds to one with a mass of 10 M_{Sun}. The left-hand end of this figure (log $M/M_{Sun} = -0.6$) corresponds to a mass $M \approx 0.25$ M_{Sun}, while the right-hand end of the figure (log $M/M_{Sun} \approx +1.7$) corresponds to $M \approx 50$ M_{Sun}. The "cloudy" region to the left shows the extent of the surface convection zone; the depth increases as one proceeds to stars with masses progressively less than the Sun's mass. The surface convection zone extends all the way to the center for stars less massive than about 0.25 M_{Sun}. Conversely, the "cloudy" regions to the lower right show the extent of the central convection zones in upper Main Sequence stars. The dashed curves labeled "0.5 L" and "0.9 L" plot the locations within which 50% and 90%, respectively, of the star's luminosity originates. The solid curves labeled "0.25 R" and "0.5 R" label the mass levels corresponding, respectively, to ¼ and ½ of the star's outer radius. (Source: Kippenhahn and Weigert 1990, p. 213; used under license number 5485520879237 from Springer Nature).

the region of high opacity therefore extends deeper into a low-mass star. Energy transport by radiation accordingly becomes less efficient, and convection extends progressively deeper than in a star of higher mass.

For stars with masses less than about one-quarter of the solar mass, the convection zone actually extends all the way down to the stellar center, so that the entire star is convective. When this occurs, the composition of the star remains uniform throughout, as it is efficiently mixed by the convection, but the hydrogen abundance gradually decreases—and the helium abundance correspondingly increases—as hydrogen is slowly converted into helium by the thermonuclear reactions in the stellar core. The effective ("surface") temperatures of these low-mass stars are far too low for any helium lines to be seen, so that the only observable clue that this is happening is that the surface pressure—as measured by the widths of the absorption lines of hydrogen and other elements—increases. The Main Sequence lifetimes of these very faint, low-mass stars are so great—many times longer than the age of the universe—that they remain essentially unchanged, apart from the very gradual reduction in their hydrogen content, from the state at which they first commenced hydrogen burning.

As the table in Appendix C shows, the central temperatures of ZAMS stars decrease as the stellar mass decreases, while the central density increases. This has two important consequences. First, as the stellar mass decreases, the temperature eventually becomes too low for H-burning thermonuclear reactions to produce enough energy to sustain the luminosity being radiated away from the stellar surface. Second, at such low temperatures and high densities, the electrons near the stellar center become degenerate, as discussed in Chapter 6. This provides sufficient pressure support to prevent the star from contracting and producing enough compressional heating to make up for the insufficient energy being produced by the residual thermonuclear reactions. These two effects mean that there is a lower limit to the mass of a H-burning Main Sequence star. This limit turns out to be about 0.1 M_{Sun}, the lowest mass listed in the table in Appendix C. Objects with masses below this limit do exist; they are called "brown dwarfs," and we discuss them further in Chapter 12.

10.2 The Upper Main Sequence

As we go up the Main Sequence to higher masses, the central temperature increases, as can be seen from the table in Appendix C. Energy production in these higher-mass stars is dominated by the CNO bi-cycle thermonuclear reactions, as indicated in Figure 10.1. At the higher central temperatures in these stars, the CNO reactions produce energy at such a prodigious rate that radiation flow is insufficient to carry it away quickly enough. For this reason, convection breaks out at the center of the star, dominating the energy transport in the core regions of these upper Main Sequence stars, as shown in Figure 10.2.

As Sir Arthur Stanley Eddington pointed out in 1926, the pressure exerted by radiation plays an increasingly important role in the structures of upper Main Sequence stars. The much more detailed ZAMS models listed in Appendix C confirm Eddington's insight. The radiation pressure, proportional to the fourth power of the absolute temperature, can easily be calculated from the central temperatures of the models. While it contributes only about 6% of the total pressure of a 10 M_{Sun} star, the percentage increases rapidly with mass, reaching 15% at 20 M_{Sun} and 36% at 60 M_{Sun}.

The increasingly strong radiation pressure exerted by the rapidly increasing luminosities of stars with higher masses is even capable of driving winds that carry mass away from the stars. Winds are hydro*dynamic*, rather than hydro*static*, phenomena and are much more complicated to describe mathematically. Because the structure of a star depends upon conditions both at its center and at its surface, the advent of radiation-driven winds significantly complicates calculations of the structures of massive stars. Consequently, the internal structure of a Main Sequence star becomes increasingly uncertain as the mass increases.

While Eddington seems not to have anticipated the existence of radiation-driven winds, he did recognize that for a star of sufficiently high luminosity, the radiation-pressure gradient alone is capable of balancing the surface gravity, making it impossible for hydrostatic models to be constructed at still-higher luminosities. Because the radiation pressure gradient is proportional to the luminosity and the surface gravity is proportional to the stellar mass, the balance between these two forces yields a critical luminosity—now called the "Eddington luminosity"—above which static stellar models cannot be constructed. The actual value of the mass limit for upper Main Sequence stars is very uncertain. However, it is certain that radiation becomes increasingly important for higher stellar masses and that there is a radiation-determined upper limit to the mass of a Main Sequence star somewhere around a few hundred solar masses.

10.3 H-burning Lifetimes

As noted previously, we now know with certainty that the Sun and other Main Sequence stars are powered by thermonuclear fusion reactions that gradually convert hydrogen into helium in their central regions. Since there is a finite supply of the nuclear fuel (hydrogen) in any given star, an obvious question is to ask how long this fuel supply can last and what happens to the star once the fuel has been exhausted.

In the Sun, the region that generates 50% of the luminosity that is ultimately radiated away from the solar surface occupies the innermost 4% of the mass, while 90% of the luminosity originates within less than the inner 30% of the mass (see Figure 10.3). As this figure also shows, the hydrogen mass fraction X_H at the center of the present Sun is today approximately half of its original value. Since the present age of the Sun is known to be about 4.5 billion years, if we were to suppose that its Main Sequence lifetime were to end when the central hydrogen abundance reaches zero, the estimated Main Sequence lifetime of the Sun would be about 9 billion years.

Alternatively, we might arbitrarily define the end of the Main Sequence lifetime as the time required to convert, say, 10% of the hydrogen in the Sun into helium. Einstein's famous equation $E = mc^2$ then shows that the thermonuclear burning of 0.1 solar masses—i.e., converting all the protons into helium nuclei in this quantity of mass—releases a total of 1.27×10^{51} ergs. If the Sun were to remain unchanged while this amount of energy is radiated away at the rate given by the present solar luminosity, the time required would be 10.5 billion years. As we shall see below in Chapter 13, these rough estimates provide a not-unreasonable approximation for the hydrogen-burning lifetime of a star.

For stars with masses progressively less than the Sun's mass, the stellar luminosity decreases very rapidly, as can be seen from the models tabulated in Appendix C. Consequently, even though the amount of hydrogen fuel available to sustain a star's power output decreases, its luminosity declines much more rapidly, and the Main Sequence lifetimes consequently increase greatly. For example, if we estimate the Main Sequence lifetime of a 0.75 M_{Sun} star as we did above for the Sun—by

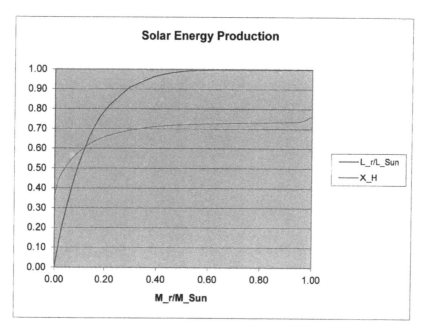

Figure 10.3. The distribution of thermonuclear energy production within the present Sun. The blue curve gives the ratio of the energy per second flowing outward through a spherical shell at the fraction of the stellar mass listed along the x-axis relative to the same quantity leaving the surface of the Sun (the total solar luminosity). The red curve lists the hydrogen mass fraction X_H (the number of grams of hydrogen per gram of matter with the bulk solar composition). The drop in the red curve toward the center of the Sun provides a measure of the amount of hydrogen converted into helium by thermonuclear reactions in the solar core from the start of hydrogen burning to the present age of the Sun. This is a plot by the author of data from Bahcall's standard solar model BS05 (AGS, OP; Bahcall et al. 2005).

assuming it to be the time required to consume 10% of the hydrogen the star contains—we find it to be a whopping 42 Gyr![2] This is about three times longer than the age of the universe! Even taking into account the approximate nature of this age estimate, this approach demonstrates that the luminosity of a lower Main Sequence star decreases so much more rapidly than does the amount of nuclear fuel available that most low-mass stars spend their entire nuclear-burning lifetimes on the Main Sequence.

Similarly, we can estimate the lifetimes of upper Main Sequence stars roughly as the time required to radiate away the energy produced by thermonuclear burning of 10% of their stellar hydrogen content. For a 2 M_{Sun} star, for example, the amount of energy released is about 2.5×10^{51} ergs. With a luminosity of 7.04×10^{34} ergs s^{-1} (from the table in Appendix C), this yields a hydrogen-burning lifetime of about 1.1 billion years, only about a quarter of the present age of the Sun. If we go further up the Main Sequence, however—say, to 20 M_{Sun}—the same type of estimate yields a total energy release of 2.5×10^{52} ergs, while the luminosity increases to 1.6×10^{38} ergs s^{-1} (again

---

[2] 1 Gyr = 1 gigayear = 1 billion years = 1000 million years.

from Appendix C), giving a hydrogen-burning lifetime of less than 5 million years. Since the hydrogen-burning lifetimes of massive stars are so much shorter than the 14-billion-year age of the universe, they cannot have been primordial but instead must have formed relatively recently. This raises the question: Where and how do stars form? We consider this in the next chapter.

References

Bahcall, J. N., Serenelli, A. M., & Basu, S. 2005, ApJ, 621, L85 The detailed data are from http://www.sns.ias.edu/~jnb?

Cox A. N. (ed) 2000, Allen's Astrophysical Quantities (New York: AIP/Springer) 471; see also p. 388–389

Hansen, C. J., & Kawaler, S. D. 1994, Stellar Interiors (New York: Springer)

Chapter 11

Where Do Stars Come From?

Around the middle of the 20th century, scientists gradually came to realize that stars are continually forming around us in the Galaxy today. One indication was provided by dynamical studies of stellar groupings containing massive stars of spectral types O and B (so-called "OB associations"). In the late 1940s, Soviet–Armenian astrophysicist Victor Ambartsumian (1908–1996) found these stellar associations to be expanding at about 10 km s^{-1}. Given their current dimensions, this yielded an expansion age of about 10 million years (Ambartsumian 1949). As Ambartsumian and his colleague L. V. Mirzoyan put it in a retrospective summary of his work, this age estimate, "is hundreds of times less than the age of most stars in the Galaxy…; therefore *the stellar associations should be regarded as young formations, while their constituent stars have been formed quite recently*. Consequently the single fact of the existence of stellar associations is evidence of the *stellar formation process that started several billion years ago in the Galaxy, and which is still continuing now*." Further, the "percentage of multiple stars, [groupings of which are] often exceedingly unstable and disintegrating (systems of the Trapezium type …), is very high in stellar associations. This fact indicates … [that] *within the association itself the stars originate not singly, but in groups*." (Ambartusmian & Mirzoyan 1982) [Authors' italics.]

The "Trapezium" is a typical OB association. It is a small cluster in the Great Nebula in the constellation Orion (Figure 11.1), which contains four very bright stars arranged in a trapezoidal shape (hence the name) as seen projected on the sky. It was originally discovered by Galileo in the 17th century, and a number of less-luminous members of the group have since been identified. The brightest stars have masses ranging from about 15 M_{Sun} (spectral type B, showing absorption lines due to neutral helium, with effective temperatures about 20,000 K and luminosities some 5000 times larger than the Sun's), up to as much as 30 M_{Sun} (spectral type O, exhibiting absorption lines produced by ionized helium, with effective temperatures between 30,000 K and 40,000 K and luminosities exceeding 200,000 times the solar

doi:10.1088/2514-3433/acce33ch11

Figure 11.1. An optical image of the Great Nebula in Orion. This is a region containing ionized hydrogen, which astronomers call an "H II region," and it contains a typical OB association—the Trapezium cluster—located in the bright region at the center of the nebula. By far the brightest is the O star θ^1 Orionis C1, which has a mass of 33 M_{Sun} and a luminosity more than 200,000 times the solar luminosity. It is providing most of the ultraviolet radiation that is gradually ionizing the nebula. This image clearly shows the cloud of gas and dust surrounding the bright stars and from which they formed. The reddish glow of the nebula is due to the Hα line of neutral hydrogen that is excited by the ultraviolet radiation from the hot stars. Credit: Courtesy of Bill Schoening/NOIRLab/NSF/AURA.

value). With a diameter of only about six parsecs and an estimated expansion age of about 3×10^5 years, this very young association is the pink middle "star" in Orion's sword. The pink glow results from strong emission from the surrounding nebular gas in the red Hα line of neutral hydrogen, which is excited by the intense ultraviolet radiation from the very hot, massive stars. Figure 11.1 shows the hot stars to be partially surrounded by dense, obscuring clouds of gas and dust. In the (astronomically) short timescale provided by its expansion age, the Trapezium stars cannot have moved further than a few parsecs from the location where they formed, demonstrating that they indeed originated within the dense cloud of gas and dust that forms the Orion Nebula. The gas and dust are currently being dissipated by the intense radiation and stellar winds produced by the massive stars. Star formation is still continuing within this dense cloud, which may be thought of as a "stellar nursery," where stars with a range of differing masses continue to be born today. About 390 pc away, the Orion Nebula is the star-formation region closest to the Sun.

Since the 1950s, astronomers have learned much more about the nature of the interstellar clouds of gas and dust from which the stars form. First, the clouds are so cold (about 10 K) that most of the atoms they contain have combined into molecules, and most of the heavy elements have condensed onto grains of interstellar dust. Since the composition of interstellar matter is mostly hydrogen, about 25% by mass of helium, and a few percent of heavy elements, the gas consists mainly of molecular hydrogen, H_2, atomic He, and molecules like HCN, H_2O, NH_3, and CO. These molecules in turn condense into icy grains. At the same time, other molecules —such as SiO, SiC_2, NaCl, etc.—condense into more refractory grains. Because the density of interstellar matter is so low—estimates (Burke & Graham-Smith 1997) range from less than 0.01 atoms cm^{-3} to more than 100 atoms cm^{-3}—the space density of the grains is lower still.

A cloud of interstellar gas and dust can remain in stable equilibrium only so long as the pressure due to the thermal energy of the gas is high enough to counterbalance the cloud's own self-gravity. Early in the 20th century, James H. Jeans[1] (1877–1946), then a graduate student at Cambridge University, considered this problem. He found that if its mass is sufficiently large, the cloud collapses under its own self-gravity. The critical point at which the thermal pressure and self-gravity just balance is called the "Jeans mass," and the corresponding radius is the "Jeans radius." The Jeans mass "depends upon the temperature and density of the cloud but is typically thousands to tens of thousands of solar masses. This coincides with the typical mass of an open cluster of stars, which is the end product of a collapsing cloud."[2]

When a cloud becomes unstable and begins to collapse, it does not do so uniformly. Instead, as the mean density of the cloud increases, the Jeans mass decreases, and smaller and smaller sub-regions become unstable and respectively begin to collapse toward their own centers. This process, called "fragmentation," was first described in an astronomical context by the late Sir Fred Hoyle (1915–2001; Hoyle 1953). The timescale for the collapse is approximately the free-fall time; that is, it is the time required for the cloud to collapse under its own self-gravity. For a cloud with a Jeans mass of about 100,000 M_{Sun}, typical of the giant molecular clouds from which stars form (Sanders et al. 1985), and a mass density of 10^{-22} g cm^{-3} (corresponding approximately to one hundred hydrogen atoms per cubic centimeter), the free-fall time is about 5 million years.

Collapse and fragmentation continue as long as the individual fragments are able to get rid of the heat produced by the gravitational compression. Because most of the heat is radiated away, the internal temperatures of the cloud fragments remain approximately constant during this early stage of the infall. Calculations show (Stahler 1988) that it occurs in an "inside-out ... fashion, with the highest-density central regions collapsing first and the outer layers remaining [temporarily] static..."

---

[1] Jeans was later knighted for his fundamental contributions to astrophysics.
[2] https://en.wikipedia.org/wiki/Star_formation

At some point the central core becomes opaque to its own radiation, and the inner regions begin to heat up. As the central temperature of a protostar reaches about 2000 K, molecular hydrogen (H_2) dissociates into atomic hydrogen, and as the temperature continues to climb, hydrogen and helium become successively ionized. As the temperature and density continue to increase, the thermal pressure exerted by the compressed gases eventually becomes great enough to counterbalance the force of gravity in the innermost core of the collapsing cloud fragment. The core is then said to be in hydrostatic equilibrium. Calculations (Shu et al. 1987) indicate that this happens when the mass of this core reaches a fraction of a percent of the Sun's mass. At this stage, the cloud fragment becomes a true "protostar."

Because of random motions in a collapsing and fragmenting gas cloud, individual protostars are born rotating. And because angular momentum is conserved, the rotation speeds up as an individual cloud fragment collapses, just as a spinning figure skater speeds up when she pulls her arms in close to her body. In general, cloud fragments possess too much angular momentum to allow them to accumulate directly onto the protostellar cores. Sometimes this causes a fragment to separate into two or more protostars, with the angular momentum of the original fragment preserved as orbital angular momenta in the resulting protostellar cluster. Binary and multiple star systems are indeed quite common, indicating that this happens frequently.

Other times, the angular momentum causes much of the cloud fragment to collapse into a thick disk of gas and dust surrounding the protostellar core. As angular momentum is dissipated by interactions within the disk, matter flows through it to accumulate on the nascent star. Larger solid particles are left behind in the disk, where they gradually accumulate into planetary systems surrounding the protostar. Once thought to be rare, extrasolar planetary systems are now known to be ubiquitous in the Galaxy.

When and how does a protostar actually become a star in hydrostatic equilibrium? In the early 1960s, the brilliant Japanese astrophysicist Chushiro Hayashi (1920–2010; see Figure 11.2) showed (Hayashi & Hōshi 1961; Hayashi et al. 1962) that protostars with effective temperatures less than a critical limit are unstable to collapse under their own self-gravity. Now called the "Hayashi limit" in his honor, this boundary is a well-defined, steep curve in the Hertzsprung–Russell diagram (the blue curve in Figure 11.3). This low-temperature region of instability (labeled the "Hayashi forbidden zone" in this figure) is precisely where protostars exist. Consequently, they continue to collapse, heat up, and become increasingly luminous until they ultimately reach their (mass-dependent) Hayashi limit, where they stabilize as fully convective pre-Main-Sequence stars. For a one-solar-mass object, this early protostellar phase lasts about 10 million years.[3] Once an object reaches this point, it continues to contract as a fully convective body, following the "Hayashi track" down to lower luminosities.

---

[3] https://en.wikipedia.org/wiki/Protostar

Figure 11.2. Chushiro Hayashi. Source: https://phys-astro.sonoma.edu/brucemedalists/chushiro-hayashi. Photo 1968, NASA GSFC, courtesy Prof. Hayashi.

Simulations show that considerable energy is released at an accretion shock front surrounding the hydrostatic core. Because this energy is absorbed and re-radiated at infrared wavelengths by dust in the infalling envelope, protostars are likely to be strong emitters of infrared radiation. Indeed, observations during the 1970s "revealed numerous embedded infrared sources and protostellar candidates in molecular clouds (Lada 1987)." One example of such a cluster of protostars appears in an infrared image of the Great Nebula in Orion (see Figure 11.4).

"At some point …an intense outflow develops and begins to arrest and reverse the infall of the outer envelope. The outflow first disrupts and removes the infalling envelope and then later the circumstellar disk ultimately revealing a reddened young star approaching or very near the ZAMS…" (Luhman et al. 2000) Because the luminosity of a protostar "is totally obscured at visual wavelengths by the dust in the infalling envelope…, it is only after the accretion process is over that the hydrostatic core is first revealed optically as a pre-main-sequence star. The core-mass–radius relation [obtained from cloud-collapse calculations] can therefore be used, in conjunction with conventional Hayashi tracks, to predict an upper envelope for pre-main-sequence stars in the H–R diagram… The envelope, or 'birthline', is in good agreement with the observed distribution of low-mass T Tauri stars in the diagram…" (Stahler 1988)

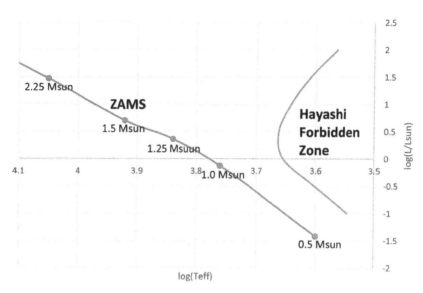

Figure 11.3. The Hayashi limit for a 1 M_{Sun} protostar. Objects in hydrostatic equilibrium cannot exist to the low-T_{eff} side of the blue curve, labeled "Hayashi Forbidden Zone." Plot by the author of data from Table 4-4 of Hayashi, Hoshi, and Sugimoto (1962). For comparison, the orange curve with masses labeled shows the low-mass end of the zero-age Main Sequence from Iben (1965), Figure 17.

Detailed model calculations for the emission from T Tauri stars show that they can be fitted very well by a combination consisting of a stellar photosphere and an accretion disk. For example, "high resolution infrared imaging of the T Tauri star HL Tau has provided strong evidence that the excess infrared emission around that object originates in a circumstellar disk... It appears that ... [such] sources represent objects around which the original infalling envelopes have been removed, but which have retained remnant star-forming material in the form of their circumstellar disks" (Lada 1987).

Once a star has reached the Hayashi track, how does it proceed from there to become a stable, hydrogen-burning Main Sequence star, and what internal changes happen along the way? By the mid-1950s, advances in computer technology were taking place that would revolutionize investigations in stellar astrophysics. At the University of California, Berkeley, astrophysicist Louis G. Henyey and his collaborators were developing a method to enable the rapid calculation of whole sequences of time-dependent stellar models. In their first paper to use the new method (Henyey et al. 1955), Henyey and his colleagues used the UNIVAC computer at the Berkeley Radiation Laboratory to compute the gravitational contraction of pre-Main-Sequence stars. In these pioneering evolutionary calculations, they studied the approach to the Zero-Age Main Sequence, where stars first ignite H-burning thermonuclear reactions. More detailed calculations of this first phase of stellar

Figure 11.4. An infrared image, taken with the Near Infrared Camera and Multi-Object Spectrometer (NICMOS) aboard the Hubble Space Telescope, of the region surrounding the Trapezium in the Great Nebula in Orion. Blue in this false-color image corresponds approximately to the near-infrared J band wavelength of 1.25 μm and red to the H band wavelength of 1.65 μm. The Trapezium is the cluster of bright stars in the center of the image, and θ^1 Orionis C1 is the brightest star at the bottom of the cluster. This image also shows a large number infrared protostars and perhaps 50 brown dwarfs. Source: https://hubblesite.org/contents/news-releases/2000/news-2000-19.html?Year=2000; image courtesy of NASA. Credit: NASA; K. L. Luhman (Harvard–Smithsonian Center for Astrophysics, Cambridge, Mass.); and G. Schneider, E. Young, G. Rieke, A. Cotera, H. Chen, M. Rieke, R. Thompson (Steward Observatory, University of Arizona, Tucson, AZ); the original single-band images are from Luhman et al. (Luhman et al. 2000).

evolution were carried out a decade later by Icko Iben, Jr (Iben 1965), then a postdoc in William A. Fowler's nuclear astrophysics group at Caltech's Kellogg Radiation Laboratory. His computations covered the final stages of gravitational contraction and the switch-over to thermonuclear hydrogen burning as the energy source for stellar models ranging in mass from 0.5 up to 15 solar masses. He found that the initial models for stars with masses less than 3 M_{Sun} were fully convective, as expected from their locations on the Hayashi tracks, while models with larger masses contained small radiative cores.

Iben described the contraction of a pre-Main-Sequence star down the Hayashi track as follows:

As the star descends in the H–R diagram, opacities in the deep interior continue to drop with rising interior temperatures until eventually the radiative temperature gradient becomes smaller than the adiabatic temperature ... [and a] radiative core develops and continues to grow monotonically with time...

In Iben's calculations for a 1 M_{Sun} model, the radiative core first develops less than a million years after the initial model. As the star steadily contracts, its radius and luminosity both decline, while the central temperature and density both increase. The luminosity reaches a minimum of about 0.4 L_{Sun} when the star is about 9 million years old. By this time, the radius has decreased from its initial value of about 3.7 R_{Sun} at the start of the calculation to about 1.2 R_{Sun}. The effective temperature has increased slightly from a bit more than 4100 K on the Hayashi track to about 4360 K, and the mass of the radiative core has grown to encompass the innermost 77% of the total stellar mass.

The minimum luminosity marks the end of a star's journey down the Hayashi track. At this stage, the 1 M_{Sun} model is still 20% larger than the present Sun, and— for some time longer—gravitational contraction continues to supply all the energy radiated away from the stellar surface.

By this point, most of the energy generated in the deep interior of the star is being transferred to the surface layers by radiation rather than by convection. This changes the star's path in the H–R diagram. Continuing gravitational contraction next leads to a rapid increase in the effective temperature and a concomitant gradual increase in the luminosity. The central temperature and density at this point have increased to about 6 million degrees Kelvin and 11 g cm^{-3}, respectively, just barely large enough for the innermost regions of the star to begin producing modest amounts of energy from hydrogen burning.

Conditions inside the 1 M_{Sun} star change dramatically by the time it reaches an age of about 25 million years. This corresponds to a relative maximum in luminosity along the final leg of the pre-Main-Sequence track before the star settles down into a prolonged phase of central hydrogen burning. By this point, the star's luminosity is provided about equally from heat produced by the slowing gravitational contraction and from increasing thermonuclear burning. The central temperature and density are now about 13 million K and 86 g cm^{-3}, respectively, and both the radius and luminosity of the star have reached 96% of the solar values. As nuclear energy generation increases, the star temporarily develops a convective core that transports energy outward from the nuclear-burning regions at the stellar center. The convective core reaches a maximum extent of about 10% of the mass before declining again to zero by the time the star reaches the Main Sequence.

The 1 M_{Sun} star finally stabilizes on the zero-age Main Sequence (ZAMS) after about 50 million years. By this point thermonuclear hydrogen-burning reactions provide all of the star's luminosity, and the central temperature and density have stabilized at 13.8 million K and 85 g cm^{-3}, respectively.

The approach to the Main Sequence is generally similar—though differing in detail—for stars with masses greater or lesser than the Sun's mass. Figure 11.5 shows

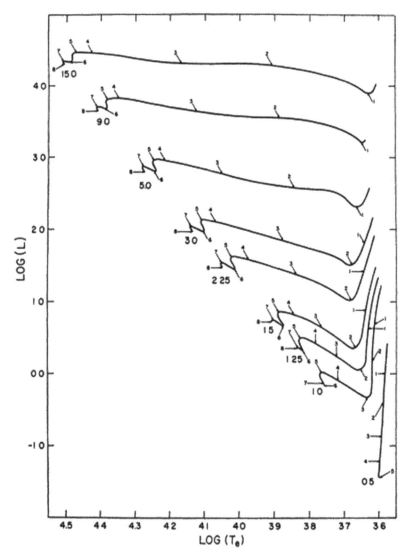

Figure 11.5. Pre-Main-Sequence evolution for stars with masses from 0.5 to 15 M_{Sun}. Source: Iben, I., Jr. 1965, ApJ. **141**, 993, Fig. 17; reproduced with permission from I. Iben, Jr.

the paths in the Hertzsprung–Russell diagram for stars ranging in mass from 0.5 to 15 solar masses, according to Iben's calculations (Iben 1965). Because the effective temperatures of stars on the Main Sequence are progressively smaller for stars with lower masses, the lower end of the ZAMS merges with the lower end of the Hayashi track for stars somewhat less massive than the Sun. This is true in particular for Iben's 0.5 M_{Sun} model, which remains fully convective all the way down to—and on —the ZAMS. It takes this low-mass star 155 million years to stabilize on the Main Sequence. Conversely, stars more massive than about 5 M_{Sun} spend little or no time on the Hayashi track. Instead, their interiors are hot enough to make radiative transport of energy more efficient than convection as soon as they have become pre-Main-Sequence stars. Because their luminosities are so high, these upper-Main-Sequence stars evolve much more rapidly to the Main Sequence. Although it takes a 1 M_{Sun} star about 50 million years to reach the Main Sequence, a 5 M_{Sun} star gets there in about 0.6 million years, and a 15 M_{Sun} star takes only about 60,000 years.

References

Ambartsumian, V. A. 1949, AZh, 26 3; see also
    Ambartsumian, V. A., & Mirzoyan, L. V. 1982, Ap&SS, 84, 317

Ambartusmian, V. A., & Mirzoyan, L. V. 1982, Ap&SS, 84, 317

Burke, B. F., & Graham-Smith, F. 1997, An Introduction to Radio Astronomy (Cambridge: Cambridge Univ. Press) 118

Hayashi, C., & Hōshi, R. 1961, PASJ, 13, 442

Hayashi, C., Hōshi, R., & Sugimoto, D. 1962, PThPS, 22, 77

Henyey, L. G., LeLevier, R., & Levée, R. D. 1955, PASP, 67, 154

Hoyle, F. 1953, ApJ, 118, 513

https://en.wikipedia.org/wiki/Protostar

https://en.wikipedia.org/wiki/Star_formation

Iben, I. Jr. 1965, ApJ, 141, 993

Lada, C. J. 1987, Peimbert, M., and Jugaku, J. (eds.), Star Forming Regions, Proc. IAU Symp #115 (Dordrecht: Reidel) 1

Luhman, K. L., Rieke, G. H., Young, E. T., et al. 2000, ApJ, 540, 1016

Sanders, D. B., Scoville, N. Z., & Solomon, P. M. 1985, ApJ, 289, 373

Shu, F. H., Adams, F. C., & Lizano, S. 1987, ARA&A, 25, 23

Stahler, S. W. 1988, PASP, 100, 1474

Chapter 12

Failed Stars: Brown Dwarfs

As a low-mass protostar contracts down its Hayashi track, the pressure and temperature at its center both rise. In its fully ionized interior, the protons act essentially like an ideal gas, with pressure increasing in proportion to the product of density and temperature. The much-less-massive electrons, however, are strongly influenced by quantum effects, which prevent more than one electron from occupying the same quantum state. The increasing density forces electrons into quantum states with higher and higher energies. In such a case, the electrons are said to be "degenerate" (see Appendix D), and the pressure exerted by the degenerate electrons increases with increasing density. If the degenerate-electron pressure equals or exceeds the ideal-gas pressure due to the protons and other massive particles, it becomes capable of supporting the object against its own self-gravity, and the object stops contracting.[1]

In 1963, astrophysicist Shiv S. Kumar—then at NASA's Institute for Space Studies in New York City—first pointed out that for objects of sufficiently low mass, electron degeneracy can halt the contraction before the central temperature ever gets high enough for sustained hydrogen burning (Kumar 1963). With the cessation of contraction, the object no longer generates compressional heat, and as the luminosity radiates energy away from the stellar surface, the internal temperature begins to decline. However, because the degenerate-electron pressure is insensitive to the temperature, the object's radius remains almost constant. The star then slowly radiates away its residual internal heat, becoming cooler and cooler and fainter and fainter for billions of years. Kumar found that the dividing line between objects that do—or do not—ignite sustained hydrogen burning occurs at a mass of about 0.08 M_{Sun}. More-massive objects become ordinary H-burning Main Sequence stars, while less-massive ones are devoid of nuclear energy sources and finish their evolution by slowly cooling off. Kumar termed the latter objects "black dwarfs."

---

[1] See Chapter 15 for a more detailed discussion of degeneracy in stars.

Because substellar objects are intermediate in mass between low-mass stars and giant planets, the mass and radius of the giant planet Jupiter provide convenient units in which to express their physical properties. Jupiter's mass is about 1000 times less than the mass of the Sun, making $0.08\ M_{Sun} = 80 \times 10^{-3}\ M_{Sun} \approx 80\ M_{Jup}$. Similarly, Jupiter's radius is about 10 times smaller than the Sun's.[2]

The similarity between substellar objects and giant planets was heightened in 1969, when infrared measurements of Jupiter and Saturn (Aumann et al. 1969) showed that they radiate, respectively, 2.7 and 2.4 times as much power as they receive from the Sun. For Jupiter, the investigators found the intrinsic luminosity to be $3 \times 10^{-7}\ L_{Sun}$ and the effective temperature to be 134 ± 4 K. Suddenly, it was possible to add the giant planets to the Hertzsprung–Russell diagram.

It did not take long for theorists to develop numerical models for the evolution of Jupiter similar to those that had been developed for stars. Important differences in the microphysics of the models included departures from the ideal-gas laws produced not only by electron degeneracy but also by electrostatic interactions and by the finite sizes of species that remain incompletely ionized at the lower temperatures of planetary interiors. Hydrogen becomes metallic at the high pressures inside the giant planets, and this also effects the material properties. The occurrence of hydrogen–helium fractionation in some models (Stevenson 1975) also needed to be addressed because it releases additional gravitational energy. Numerous molecular species—visibly present in Jupiter's banded cloud decks—affect the opacity of the surface layers, which are critically important for fully convective models, as pointed out by Hayashi in the early 1960s.

A pioneering evolutionary calculation incorporating many of these effects was carried out in the mid-1970s by Harold C. (Hal) Graboske, Jr. and his collaborators (Graboske et al. 1975). They found that Jupiter evolves through two distinct phases: the first is an "early contraction phase, behaving like a typical low-mass pre-main-sequence object, … [which] has high luminosities … [and] internal temperatures…" The maximum central temperature in their Jupiter calculation reached 51,400 K before beginning a long, slow decline. "The second phase is an approach to a degenerate dwarf[3] cooling curve, which gives excellent agreement with the observed radius and luminosity of Jupiter." However, the model calculations reached these values after only 2.6 billion years—substantially shorter than the known 4.5 billion-year age of the solar system—and the authors speculated about a number of factors, such as H/He phase separation, that might account for this difference.

That same year, in her Berkeley Ph.D. thesis (Tarter 1975), astrophysicist Jill Tarter coined the now-accepted name "brown dwarf" for objects with masses less than the minimum necessary for hydrogen burning (Tarter 1986). Her thesis included an exploration of "what, if any, observational constraints could be placed on the assertion that a population of brown dwarfs could account for the 'missing

---

[2] The actual values are $M_{Jup} = 1.8987 \times 10^{30}$ g $= M_{Sun}/1047.6$ and $R_{Jup} = 7.1492 \times 10^9$ cm $= R_{Sun}/9.7285$; see Cox 2000, pp. 12 and 295.

[3] A "degenerate dwarf"—more commonly called a "white dwarf" (see Chapter 15)—is an object supported entirely by the pressure exerted by the degenerate electrons in its interior.

mass' within clusters of galaxies, massive haloes surrounding spiral galaxies and in the local solar neighborhood" (Tarter 1986). At the time, the idea that the "missing mass" or "dark matter" might consist of very faint stars was considered to be a real possibility. (It has since been discounted.) Tarter subsequently recounted the origin of the term "brown dwarf" used in her Ph.D. thesis (Tarter 1986):

> The fourth chapter .... entitled 'Brown Dwarf Stars and How They Grow Old'... marks the first written use of the term brown dwarf... [It arose in] a conversation between Joe Silk [her thesis advisor] and myself in the library on the 5th floor of Campbell Hall soon after I began to try and answer the question Joe had originally posed. The terms white dwarf, red dwarf and black dwarf all existed in the astronomical vocabulary. But they referred specifically to the end products of low mass stellar evolution, to low mass main sequence stars and to the end products of white dwarf cooling with a mean molecular weight that is characteristic of billions of years of nucleosynthesis. It was obvious that we needed a color to describe these dwarfs that was between red and black. I proposed brown and Joe objected that brown was not a color. As I already understood that the inadequacies of existing opacity tables and numerical codes would prevent us from deriving a meaningful color temperature for these objects, brown, a non-color, seemed to me to fit our needs precisely.

By the middle of the 1980s, theorists had developed reasonably detailed models for low-mass stars and substellar objects evolving down the Hayashi track (D'Antona 1986). For objects with masses equal to or greater than 0.085 M_{Sun}, the temperature ultimately stabilizes, indicating that they have become true hydrogen-burning stars on the very-low-mass end of the Main Sequence. Conversely, for objects with masses less than or equal to 0.075 M_{Sun} the central temperature reaches a maximum of less than 3 million K after about 100 million years and then declines continuously, even after billions of years of evolution. Thus, protostars with masses $M \geqslant 0.085 \ M_{Sun}$ become true stars, while those with $M \leqslant 0.075 \ M_{Sun}$ become brown dwarfs.

A brief period of thermonuclear burning of primordial deuterium (D = ^{2}H), which occurs at a much lower central temperature than do the conventional reactions of the proton–proton chain, causes a temporary delay in the final cooling. Because the primordial abundance of deuterium is so low [about 150 deuterium nuclei per million protons (= ^{1}H)], insufficient energy is released during D-burning to do more than create a temporary pause in the gravitational contraction down the Hayashi track. Figure 12.1 shows the evolutionary tracks of low-mass stars and brown dwarfs in the H–R diagram (D'Antona & Mazzitelli 1985). Each curve represents the evolutionary path of a model of given mass (in solar units). The locations at which deuterium burning occurs are marked on each track by small squares, while the point at which each low-mass star stabilizes on the Main Sequence is marked by a small triangle. The ages of the brown dwarf models are marked by

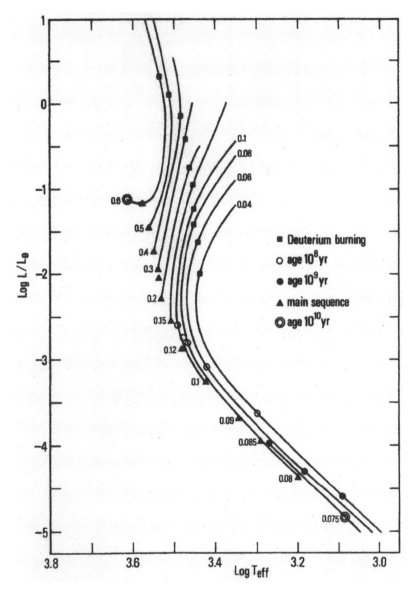

Figure 12.1. Hayashi tracks for low-mass stars and brown dwarfs in the theorists' H–R diagram. Source: D'Antona & Mazzitelli (1985); image used with permission from F. D'Antona. The vertical axis gives the logarithm of the luminosity in solar units, and the horizontal axis gives the logarithm of the effective temperature in degrees Kelvin. The present Sun would be plotted in this diagram at log L/L_{Sun} = 0 and log T_{eff} = 3.76. Each curve is labelled by the mass of the model in units of the solar mass. See text for details.

circles of different types, as listed in the legend. Note that the effective ("surface") temperatures of the billion-year-old models range from less than 1900 K (for the 0.075 M_{Sun} model) down to slightly more than 1000 K (for the 0.040 M_{Sun} model), and still older models are even cooler. At these temperatures, the peak radiation from the surface of such a substellar object lies well into the infrared region of the spectrum.

Through the 1960s, 70s, and 80s, astronomers conducted observational searches for brown dwarfs using a variety of techniques (Kafatos 1986; Reid & Hawley 2005a). The methods included astrometric measurements for low-mass companions of known faint stars, infrared surveys, precision measurements of radial velocities (precursors of the methods being used today with great success to locate planetary companions of nearby stars), so-called "speckle interferometry," and other techniques. The common goal was to search for star-like objects that were extremely faint or had very low mass or were exceptionally cool and red. While the searches did turn up a small number of such objects, by the middle of the 1980s, none had been definitively identified as being lower in mass than the minimum necessary for sustained hydrogen burning on the Main Sequence.

The situation changed dramatically in 1995, however, with the discovery of Gl 229B (see Figure 12.2) (Nakajima et al. 1995), a faint infrared companion to the low-mass star Gliese 229, the two hundred and twenty-ninth object in a catalogue of nearby stars compiled decades earlier by astronomer Wilhelm Gliese (1915–1993). This marked a turning point in brown dwarf research. "Detailed comparison of the

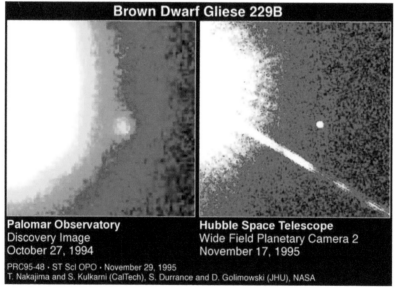

Figure 12.2. The discovery image of the brown dwarf Gliese 229B (left panel) and the confirmation image obtained with the Wide Field and Planetary Camera 2 aboard the Hubble Space Telescope (HST). The discovery image was obtained on 1994 October 27 with the 60-inch telescope at the Mt. Palomar Observatory, and the HST image was obtained on 1995 November 17. The greatly overexposed parts of the images are the faint, low-mass Main Sequence star Gliese 229, and the smaller yellow blob in each image is the brown dwarf companion. A spectrum of Gl 229B obtained with the 200-inch (5-meter) Hale Telescope at the Mt. Palomar Observatory confirmed that the spectral features of this object are in fact those of a brown dwarf. Source: Nakajima et al. (1995); http://starchild.gsfc.nasa.gov/Images/StarChild/questions/brown_dwarf.jpg; image courtesy of NASA.

spectral energy distribution of Gl 229B with theoretical models leads to a temperature of 960 ± 70 K;... this is far too cool for a hydrogen-burning star and, like the luminosity [about 6×10^{-6} L_{Sun}], clearly identifies Gl 229B as a brown dwarf.... Gl 229B probably has a mass between 30 and 60 Jovian masses (Reid & Hawley 2005b)."

The discovery of Gl 229B both proved that brown dwarfs really do exist and stimulated searches to discover more of them. Over the following decade and a half, a number of surveys found hundreds of brown dwarfs. Over the four-year period from 1997 to 2001, for example, the Two Micron All-Sky Survey (2MASS) compiled a Point Source Catalog containing 471 million objects, including more than 100 brown dwarfs (Skrutskie et al. 2006). The ground-based surveys were complemented by NASA's *Wide-Field Infrared Survey Explorer* (*WISE*), which discovered hundreds more brown dwarfs during its two-year mission (Kirkpatrick et al. 2011; Cushing et al. 2011). By 2015, these and other surveys at infrared wavelengths—where the peak of the spectrum occurs in these lower-temperature objects—had identified more than 2800 brown dwarfs (https://en.wikipedia.org/wiki/List_of_brown_dwarfs).

Because of their low temperatures, brown-dwarf spectra contain features not seen in ordinary low-mass stars. The coolest Main Sequence stars had previously been categorized as spectroscopic class M. Now, for the first time in a century, astronomers felt the need to extend the classification system to include much cooler objects. Spectroscopist J. Davy Kirkpatrick described the situation in the following way Kirkpatrick 2001:

> Which letters are left for use as possible new spectral classes? ...the best remaining letters are 'H', 'L', 'T', and 'Y' after previously-used and confusing letters are eliminated.... If either 'H' or 'L' were used, it would be a letter originally used but later dropped from the Harvard classification system... Neither 'T' nor 'Y' has been previously used as a spectroscopic class... As even more objects ... were being discovered between 1997 and 1998 by the Two Micron All Sky Survey (2MASS)..., I held discussions with colleagues in the field and with members of IAU Commission 45 on Stellar Classification to address the need for new spectral types to follow 'M'. The vast majority felt that new designations separate from 'M' were indeed warranted based on the new spectra being collected. Of the preceding four letter choices, 'L' was the overwhelming favorite to follow 'M'. It was decided that beyond this, one would proceed alphabetically down the list of remaining letters. Following 'L' would be 'T', to designate spectra such as Gl 229B with methane absorption at 2.2μm. Following 'T' would be 'Y' if needed later ...

One of the main spectroscopic characteristics of the L dwarfs (Reid & Hawley 2005c) is the decreasing strength of the TiO bands, which characterize spectral class M at the bottom of the Main Sequence. In the L dwarfs the dominant molecular features are instead due to various metal hydrides (CaH, MgH, FeH, etc.), and lines produced by the neutral alkali metals (Na, K, etc.) are also prominent. In 2000, Gilles Chabrier and his colleagues found (Chabrier et al. 2000) that dust forms in the atmospheres of models with effective temperatures less than 2800 K. And in models with effective temperatures less than about 1300 K to 1400 K, methane absorption causes the models to develop very "blue" infrared colors. This turns out to play a major observational role in the transition from L to T spectral types as a brown dwarf cools. In addition, Chabrier et al. found that the most massive brown dwarfs develop conductive cores after about 2 billion years of cooling, when electron degeneracy becomes important. The L dwarfs span the range of effective temperatures from approximately 2200 K down to about 1250 K (Kirkpatrick et al. 2021; see Figure 12.3).

The prototype for the L dwarfs is GD 165B, a faint companion to the white dwarf star GD 165. It was actually discovered in 1988 (Becklin & Zuckerman 1988), seven years before Gl 229B. Although Eric Becklin and Ben Zuckerman had identified it as a brown-dwarf *candidate*, there were enough uncertainties about its true nature (Kirkpatrick et al. 1993), that its status remained unresolved for more than a decade. Ultimately, high-signal-to-noise spectra (Kirkpatrick et al. 1999) revealed broad absorption by neutral potassium and metal hydride bands, which have come to be recognized as characteristics of spectral type L. Comparisons with theoretical models indicate that it lies very close to the border between the minimum mass for a Main Sequence star and a substellar object, explaining why it proved to be so difficult to identify it as a true brown dwarf. For this prototypical L dwarf (Kirkpatrick et al. 1999), fitting the observed spectrum to the emergent spectrum from a NextGen model atmosphere calculation by France Allard and her colleagues, which includes dust grains formed in the atmosphere as well as atomic and molecular species, yielded $T_{\text{eff}} = 1900 \pm 100$ K, $\log g = 5.0 \pm 0.5$, and an age somewhere in the range from 1.2 to 5.5 Gyr.[4]

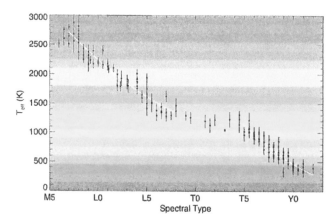

Figure 12.3. The correlation between the effective temperatures and spectral types of brown dwarfs. The rainbow color scale represents higher temperatures (violet) down to cooler temperatures (red). Source: Kirkpatrick et al. (2021); image used with permission from J. D. Kirkpatrick.

---

[4] 1 Gyr = 1 billion years.

In cooler brown dwarfs like Gl 229B, methane bands are the dominant spectral features; they do not appear in hotter objects. For a time, such objects were therefore termed "methane dwarfs," but spectroscopists ultimately defined them as constituting the new spectral class T, with Gl 229B being the prototypical member of this class. Subsequent brown dwarf model atmospheres yielded two best fits to the spectrum of Gl 229B: one with $T_{eff} = 800$ K, $\log g = 4.75$, and $M = 0.028\ M_{Sun}$, and the other with $T_{eff} = 850$ K and $\log g = 5.0$ (Nakajima et al. 2015). However, subsequent measurements of the dynamical mass of Gl229B yielded $M = 71.4 \pm 0.6\ M_{Jup}$ (Brandt et al. 2021). The disagreement with the mass obtained from the model atmosphere calculations suggests that Gl 229B may actually be an unresolved brown-dwarf binary.

In addition to methane, the spectra of T dwarfs contain (Reid & Hawley 2005d) strong lines of neutral sodium and potassium. Absorption due to the metal hydrides weakens as the effective temperature decreases, while molecular absorption by H_2O becomes stronger. In 2006, Adam Burgasser and his colleagues presented a unified spectral classification scheme for T dwarfs based on the strengths of the H_2O and CH_4 bands, together with a compendium of updated spectral classifications for all then-known T dwarfs (Burgasser et al. 2006). The effective temperatures of the T dwarfs plateau around 1250 K between spectral classes T0 and T5 before continuing to decline on down to about 500 K (see Figure 12.3). We shall return to the transition between L and T dwarfs below.

In 2007, using the infrared capabilities of the Spitzer Space Telescope, Kevin Luhman and his colleagues discovered T dwarf companions to two nearby Main Sequence stars: the T2.5 brown dwarf HN Peg B and the T7.5 object HD 3651 B (Luhman et al. 2007). Comparisons with theoretical evolutionary models yielded the masses $M = 0.021 \pm 0.009\ M_{Sun}$ for HN Peg B and $M = 0.051 \pm 0.014\ M_{Sun}$ for HD 3651 B. Because the effective temperature of HN Peg B appears to be significantly lower than the values for other T dwarfs with similar spectral types, the authors concluded that the transition between L dwarfs and T dwarfs appears to depend on the brown-dwarf surface gravity. The spectrum of the T2.5 brown dwarf HN PegB exhibits molecular bands (see Figure 12.4), and models with condensates or clouds fit the spectrum better (Suárez et al. 2021).

The even cooler spectral class Y was needed barely a decade into the 21st century. Using data from *WISE*, astronomers identified seven ultracool dwarfs with deep absorption bands of H_2O and CH_4 (Cushing et al. 2011). At least two also displayed absorption bands ascribed to ammonia (NH_3), which had come to be regarded as a trigger for the new spectral class. The brown dwarf WISE J182831.08+265037.8 became the prototype for spectral class Y. The Y dwarfs are all cooler than about 500 K, and the coolest discovered to date—WISE 1828+2650—has a surface temperature less than 300 K, which is less than 80 °F! (Cushing et al. 2011)

Theorists have steadily improved models for the emergent spectra of brown dwarfs and have used them to extract physical properties from the observations (see the examples in Appendix E). In 2006, Didier Saumon and his coworkers constructed model atmospheres containing a variety of molecular species including

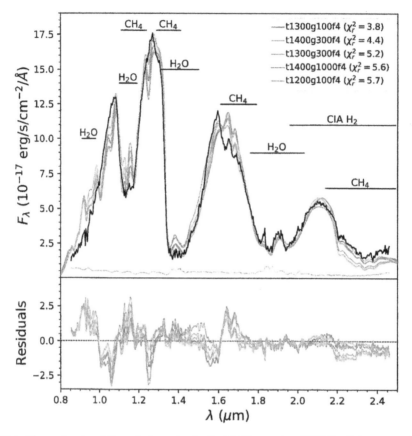

Figure 12.4. Observational and theoretical spectra for the T2.5 brown dwarf HN Peg B (upper panel), from Suárez et al. (2021); image used with permission from G. Suárez. The black line shows the observed spectrum, and the various colored curves show the spectra for various model atmospheres from Saumon and Marley (2008). The positions of various H_2O and CH_4 absorption bands are shown, as is the region dominated by collision-induced absorption by molecular hydrogen (H_2, labelled "CIA H_2"). The bottom panel shows the residual differences between the models and the observed spectrum.

NH_3 and compared the results with the observed spectrum of the T7.5 dwarf Gl 570D. They obtained T_{eff} = 800–820 K, log g = 5.09–5.23, and log (L/L_{Sun}) = −5.525 to −5.551 (Saumon et al. 2006). Two years later, he and Mark Marley showed that the transition from L dwarfs to T dwarfs corresponds to a transition from objects with cloudy atmospheres (L dwarfs) to cloudless atmosphere (T dwarfs; Saumon & Marley 2008). These evolutionary models also correctly predict the observed spike in the space density of brown dwarfs across the L–T boundary (Kirkpatrick et al. 2021).

In the half century since brown dwarfs were first suggested, they have thus gone from being a gleam in a theoretician's eye to becoming a well-established subfield of modern astronomy and astrophysics.

References

Aumann, H. H., Gillespie, C. M. Jr., & Low, F. J. 1969, ApJ, 157, L69

Becklin, E. E., & Zuckerman, B. 1988, Natur, 336, 656

Brandt, G. M., et al. 2021, AJ, 162, 301

Burgasser, A., et al. 2006, ApJ, 637, 1067

Chabrier, G., Baraffe, I., Allard, F., & Hauschildt, P. 2000, ApJ, 542, 464

Cushing, M. C., et al. 2011, ApJ, 743, 50

D'Antona, F. 1986, in Kafatos, M. C., et al. 1986, Astrophysics of Brown Dwarfs (Cambridge: Cambridge Univ. Press) 148

D'Antona, F., & Mazzitelli, I. 1985, ApJ, 296, 502

Graboske, H. C. Jr., Pollack, J. B., Grossman, A. S., & Olness, R. J. 1975, ApJ, 199, 265

https://en.wikipedia.org/wiki/List_of_brown_dwarfs

Kafatos, M. C., et al. 1986, Astrophysics of Brown Dwarfs (Cambridge: Cambridge Univ. Press) 3 ff

Kirkpatrick, J. D. 2001, in Jones, H. R. A., & Steele, I. A. (eds.) 2001, Ultracool Dwarfs (Heidelberg: Springer) 139

See also §5 and Table 5 in Kirkpatrick, J. D., et al. 1999, ApJ, 519, 802

Kirkpatrick, J. D., et al. 1999, ApJ, 519, 834

Kirkpatrick, J. D., et al. 2011, ApJS, 197, 19

Kirkpatrick, J. D., et al. 2021, ApJS, 253, 7

Kirkpatrick, J. D., Henry, T. J., & Liebert, J. 1993, ApJ, 406, 701

Kumar, S. S. 1963, ApJ, 137, 1121

Luhman, K., et al. 2007, ApJ, 654, 570

Nakajima, T., et al. 2015, AJ, 150, 53

Nakajima, T. 1995, Natur, 378, 463

Reid, I. N., & Hawley, S. L. 2005a, New Light on Dark Stars (Chichester: Springer/Praxis Publishing) 244

Reid, I. N., & Hawley, S. L. 2005b, New Light on Dark Stars (Chichester: Springer/Praxis Publishing) 250

Reid, I. N., & Hawley, S. L. 2005c, New Light on Dark Stars (Chichester: Springer/Praxis Publishing) 54

Reid, I. N., & Hawley, S. L. 2005d, New Light on Dark Stars (Chichester: Springer/Praxis Publishing) 55

Saumon, D., et al. 2006, ApJ, 647, 552

Saumon, D., & Marley, M. S. 2008, ApJ, 689, 1327

Skrutskie, M. F., et al. 2006, AJ, 131, 1163

Stevenson, D. J. 1975, PhRv, 12B, 3999

Suárez, G., et al. 2021, ApJ, 920, 99 This paper, which includes both infrared observations and models, concludes that young (0.1–0.3 Gyr) T dwarfs have effective temperatures about 140 K lower than, radii about 20% larger than, and luminosities similar to field brown dwarfs older than about 1 Gyr with similar spectral types.

Tarter, J. C. 1975, The Interactions of Gas and Galaxies within Galaxy Clusters, (Univ. California) PhD thesis

Tarter, J. C. 1986, in

Kafatos, M. C., et al. 1986, Astrophysics of Brown Dwarfs (Cambridge: Cambridge Univ. Press) 121

Part IV

Why and How Stars Evolve and Die

We have seen that stars exhaust their hydrogen thermonuclear fuel in shorter and shorter times as we progress up the Main Sequence to more massive stars. Something must therefore change inside a star in order to provide the energy needed to sustain the luminosity being radiated away from its surface. Such changes cause the structure and appearance of the star to evolve. In this section of this book, we describe the development of astrophysicists' current understanding of stellar evolution. It turns out that this also enables us to understand the nature of the various types of stars that populate regions of the Hertzsprung–Russell diagram other than the now well-understood Main Sequence.

Chapter 13

What Happens When a Low-Mass Star Exhausts Its Nuclear Fuel?

In Chapter 10, we found that once stars form, those with sufficiently low masses and luminosities continue to burn hydrogen on the Main Sequence during the entire age of the universe. However, a Main Sequence star with a H-burning lifetime less than that age ultimately finds itself in a dilemma: energy continues to flow out of its central regions at the rate established by thermonuclear burning, but when the hydrogen is exhausted from its deep interior it is no longer able to generate sufficient energy through hydrogen-burning reactions. When this occurs, the now-helium-rich core of the star begins to contract. Then, just as happened when the star initially contracted to the Main Sequence, part of the gravitational energy released by the contracting helium core goes into supplying the energy that continues to flow out of the star's central regions, while the rest goes into increasing the core temperature.

13.1 From H Core Burning to H Shell Burning

There is an important difference between pre- and post-Main Sequence evolution, however: pre-Main Sequence contraction involves the entire star, while post-Main Sequence evolution involves contraction only of the helium core. Hydrogen continues to burn in a shell surrounding the contracting helium core. If the outer layers were to contract at the same rate as the core, the temperature of the H-burning shell would increase too much, and it would generate energy too rapidly. Instead, as the helium core contracts, the outer, non-burning layers of the star expand. The H-burning shell surrounding the He core performs a delicate balancing act, with its thickness and temperature adjusting to produce exactly the amount of energy needed to maintain the star in a state of quasi-static equilibrium.

13.2 Up the Red Giant Branch

As a star expands when it leaves the Main Sequence, it moves to lower effective temperatures in the Hertzsprung–Russell diagram (thus becoming redder), and its subsurface convection zone grows inward toward the stellar core. As core contraction and envelope expansion proceed, a post-Main-Sequence star eventually reaches the Hayashi limit. As discussed in Chapter 11, objects in hydrostatic equilibrium cannot exist with effective temperatures cooler than this limit. Consequently, as the core continues to contract and the envelope expands, transforming the star into a giant—with the effective temperature now fixed at the Hayashi limit—the stellar luminosity increases, and the star "climbs up the red-giant branch."

During the early 1950s, Fred Hoyle (1915–2001), Martin Schwarzschild, and their coworkers laboriously followed the evolution of a star up the red-giant branch (RGB) to the point of helium ignition at the "red-giant tip" (Hoyle & Schwarzschild 1955). At the time, they had to calculate by hand the distributions of physical quantities throughout the star, for each of perhaps a hundred spherical shells making up a single stellar model and for each of several scores of models making up the evolutionary track. A decade later, Schwarzschild and his colleagues at Princeton (Schwarzschild & Selberg 1962) employed the new computerized approach to stellar evolution pioneered in the mid-1950s by astrophysicists at Berkeley to follow a 1.3 M_{Sun} model from the point at which it exhausts its hydrogen fuel at the center up to the point at which He burning is about to begin, marking the end of the star's journey up the RGB.

At about the same time, Icko Iben Jr.—then a postdoc in William A. Fowler's nuclear astrophysics group at Caltech—carried out similar computer modeling. For stars with masses ranging from 3 to 15 M_{Sun}, he was able to follow the evolution all the way from the Main Sequence through core He burning and beyond, as described in the following chapters. For models with 1.0 and 1.5 M_{Sun}, however, like Schwarzschild he was only able to follow the evolution up to the point of He ignition (Iben 1967), for reasons discussed below. For models with masses of 1.0 M_{Sun}, 1.25 M_{Sun}, and 1.5 M_{Sun}, he found the time required to reach the red-giant tip to be about 10.9, 4.5, and 2.3 billion years, respectively.

A few years later, Pierre Demarque and John Mengel (1945–2007) at Yale University followed the evolution of a number of low-mass stellar models all the way to the start of He burning at the red giant tip (Demarque & Mengel 1971). At this point, the luminosity of a 0.85 M_{Sun} model, for example, had increased to about 1200 times the solar luminosity, even though the effective temperature was only about 4400 K—only about three-fourths the surface temperature of the Sun—and the stellar radius had expanded to about 60 times the solar radius, more than 70% of the distance from the Sun to the orbit of the planet Mercury! At that point, almost half the mass of the entire star—down to a radius only a few percent of the Sun's radius—was undergoing boiling convection in order to transport the great amount of energy being generated by the H-burning shell out to the stellar surface. The density of the gases in this outer part of the star was very low, less dense than a cloud in Earth's atmosphere. The temperature gradually increased inward through the star, reaching about 5 million Kelvin at the H-burning shell located only about 0.03 R_{Sun} from the stellar center—just three times the radius of the Earth! This small radius nevertheless contained a 0.46 M_{Sun} He core with a central temperature of almost 90 million Kelvin and central density of about 5.4×10^5 g cm^{-3}. The latter is so dense that a chunk of matter about the size of a sugar cube would weigh approximately 1200 pounds!

13.3 Ignition of He burning in Low-mass Stars

As the core of a red giant star contracts, at some point it becomes hot enough to ignite He burning. The obvious reaction to consider, in analogy with the first reaction in the proton–proton chain, is

$$^4\text{He} + {}^4\text{He} \rightarrow {}^8\text{Be}.$$

The problem with this, however, is that the ground-state energy of the ^{8}Be nucleus is 92 keV higher in energy[1] than the combined mass–energy of two alpha particles. Consequently, the ^{8}Be nucleus decays almost immediately back into two alpha particles:

$$^8\text{Be} \rightarrow {}^4\text{He} + {}^4\text{He}.$$

Up until 1952, astrophysicists had regarded the instability of the ^{8}Be nucleus as a major stumbling block, preventing the formation of heavier elements. However, in a short (three-page) paper (Salpeter 1952) in that year, Edwin E. Salpeter (1925–2008; Figure 13.1) pointed out that the finite—though very short—lifetime of the ^{8}Be nucleus means that in thermal equilibrium at a temperature near 200 million degrees K, a tiny concentration of ^{8}Be nuclei will be present in the "sea" of ^{4}He nuclei. A further reaction can then take place:

$$^4\text{He} + {}^8\text{Be} \rightarrow {}^{12}\text{C}.$$

Because the ground state of the ^{12}C nucleus lies 7.366 MeV below the combined rest-mass energies of the ^{4}He and ^{8}Be nuclei, this reaction proceeds spontaneously in the direction indicated by the arrow, releasing energy (in the form of a gamma-ray photon) in the process.

However, the rate of this direct reaction to the ^{12}C ground state is much too slow to produce a significant amount of this nuclide, despite ^{12}C being relatively abundant in nature. This fact led British astrophysicist Fred Hoyle (Figure 13.2) to speculate that the ^{12}C nucleus must have an excited state about 7.65 MeV above the ground state that is in resonance with the combined energy of the three alpha particles (^{4}He nuclei), the resonance greatly increasing the reaction rate. Fowler and his colleagues at Caltech soon confirmed that Hoyle's postulated energy level actually does exist! (Cook et al. 1957)

---

[1] An electron volt, abbreviated "eV," is a convenient measure of energy to use in connection with electronic levels in atoms. One million electron volts, or 1 MeV, which equals 1.602×10^{-6} ergs, is a more convenient unit for use in connection with nuclear interactions. One thousandth of an MeV equals 1 keV, one kilo electron volt.

Figure 13.1. Edwin E. Salpeter. Credit: AIP Emilio Segrè Visual Archives, John Irwin Slide Collection. Copyright: American Institute of Physics.

Because these reactions effectively combine three alpha particles into one ^{12}C nucleus, they are collectively termed the "3α" or "triple-alpha" reaction. Subsequent α captures convert some of the ^{12}C nuclei into ^{16}O, and then into ^{20}Ne, and so forth, albeit at decreasing rates. The 3α reaction is thus a source of thermonuclear energy that can "turn on" when gravitational contraction compresses the He core of a post-Main-Sequence star to make it hot enough.

In stars with initial Main Sequence masses less than about 2.3 M_{Sun}, calculations had shown that the helium core becomes partially degenerate[2] before helium ignition occurs. At that point, the central density of the star exceeds 100,000 g cm^{-3} and the central temperature exceeds 120 million degrees Kelvin. These are, respectively, about a thousand times greater than the density at the center of the Sun and about 10 times greater than the temperature there. Because the pressure of degenerate matter

---

[2] See Appendix D for a more detailed discussion of electron degeneracy.

Figure 13.2. Fred Hoyle at Sloan Laboratory at Caltech in 1967. Credit: AIP Emilio Segrè Visual Archives, Clayton Collection. Copyright: American Institute of Physics.

is insensitive to temperature, the onset of He burning causes a rapid rise in temperature until the core becomes hot enough to lift the degeneracy. In 1962, Schwarzschild and Härm (Schwarzschild & Härm 1962) used an IBM 650 computer to construct 70 models to follow the evolution of the core of a 1.3 M_{Sun} model through the ignition of He burning.[3] They found that when the temperature increases enough to lift the degeneracy, the pressure begins to increase, causing the core to expand and slowing the He burning reactions. This defines the tip of the red giant branch. The expansion happens so rapidly that Schwarzschild and Härm

---

[3] Because their calculations included only the helium core of the star, Schwarzschild and Härm could not provide reliable stellar radii, nor could they produce evolutionary tracks.

referred to He ignition in the cores of these low-mass stars as the "He flash." While the timescale for significant changes to occur had been many millions of years when the star first began to leave the Main Sequence, it drops to mere seconds as the star approaches the tip of the red-giant branch.

Because of the very short evolutionary timescale at the peak of the He flash, astrophysicists wondered whether He ignition might be sufficiently violent to eject matter from the star in a hydrodynamic event, as in fact *does* happen during the explosive ignition of hydrogen shell burning in a nova outburst (see Chapter 21). A number of calculations in the 1970s and 1980s did suggest that this might actually be the case. However, sophisticated hydrodynamic simulations carried out late in the first decade of the 21st century (Dearborn et al. 2006) appear to have put this idea to rest. While very rapid by comparison with the normal timescales for stellar evolution (often millions to billions of years), the He flash nevertheless is *not* an explosive hydrodynamic event. However, at the high luminosities and low surface gravities near the tip of the red-giant branch (RGB), the star *does* lose a significant amount of mass, presumably through some type of stellar wind (see below). Consequently, the star's mass becomes appreciably less than its initial Main Sequence mass through the subsequent phases of stellar evolution.

13.4 He Core Burning: The Horizontal Branch

In 1966, astrophysicist John Faulkner had an epiphany. At the time, the British-born, Cambridge-educated Faulkner was a Research Fellow in William A. Fowler's nuclear astrophysics group at Caltech. In one of those rare leaps of imagination that can lead to a significant advance in a field, Faulkner pointed out that a stellar model with both a H-burning shell and a He-burning core provides an explanation for the previously mysterious class of stars that populate the so-called "horizontal branch" (HB) in the Hertzsprung–Russell (or "color–magnitude") diagram of a globular star cluster (Faulkner 1966).

Globular clusters are collections of some hundreds of thousands of stars, and they are so old that any initial irregularities in the shape of the cluster have long since been smoothed out, leaving it very nearly spherical in shape. The lower panel in Figure 13.3 shows a photograph of the globular cluster M5,[4] while the upper panel is a form of the H–R diagram for this cluster, called the "color–magnitude diagram" or "cluster diagram." Each point in the color–magnitude diagram represents the apparent magnitude V of a single star, plotted against its $B-V$ color index. Because

---

[4] In 1771, the famous French comet-hunter Charles Messier, tired of "wasting time" on non-cometary cloudy patches in the heavens with the relatively poor-quality optical instruments available to him, published a catalog listing a number of these objects (see Pannekoek 1961, p. 312). As telescope quality improved, astronomers gradually found that Messier's troublesome nebulosities included a wide range of astronomical objects. For example, the first item in his catalog, M1, we know today as the Crab Nebula, a supernova remnant; M5 proved to be a globular star cluster; M31 is the Andromeda galaxy; and M42 and 43 are the Great Nebula in Orion, the middle "star" in Orion's sword. A complete list of the Messier objects is given by Cox (2000, pp. 674–675). M5 is also known as NGC 5904, because it is listed as object number 5904 in the "New General Catalogue," which was originally compiled in the latter part of the 19th century under the auspices of Britain's Royal Astronomical Society.

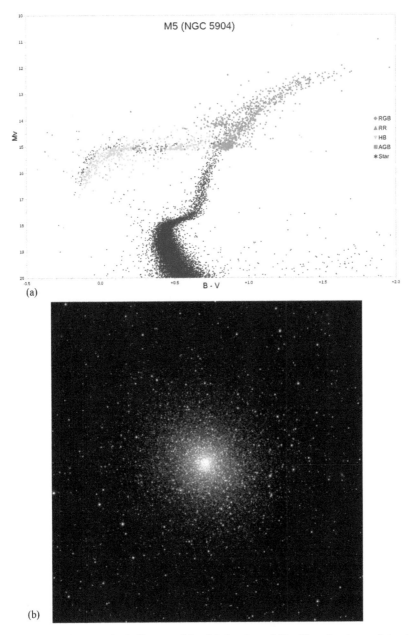

Figure 13.3. (a) The color–magnitude diagram of the globular cluster M5, with various types of stars indicated by color coding. Author: Lithopsian, CC BY-SA 4.0. Source: https://commons.wikimedia.org/wiki/File: M5_colour_magnitude_diagram.png. (b) An image of the cluster. Credit: Hillary Mathis, REU Progam/ NOIRLab/NSF/AURA. Source: noirlab.edu/public/media/archive/images/large/noao-m5.jpg.

all the stars in the cluster lie at approximately the same distance—M5 is 9200 parsecs distant, while its diameter is only 12 parsecs (Allen 1955)—the difference between the apparent magnitude V and the absolute magnitude M_V is almost exactly the same additive constant for every star in the cluster. The color–magnitude diagram can thus be converted into an H–R diagram by a simple vertical shift. The relative positions of all the stars are preserved. Stars with larger values of $B-V$ are redder, while those with smaller—or negative—values are bluer. In the cluster diagram for M5, those stars fainter than about $V = +18$ (larger values of V correspond to increasingly faint stars) and with $B-V$ colors ranging from about +0.4 to about +0.8 are still on the H-burning Main Sequence. The population of stars extending from the top of the Main Sequence near $V \approx +18$ and $B-V \approx +0.4$ to the upper right in the diagram defines the red-giant branch (RGB) for the cluster. The red-giant tip is located at $V \approx +12$ and $B-V \approx +1.6$. The group of stars extending from $V \approx +15$ near the red-giant branch downward to the left (bluer colors), ending near $V \approx +16.5$ and $B-V \approx -0.3$, is called the "Horizontal Branch" (HB). For comparison, the color of the Sun, with an effective temperature of 5800 K, is $B-V = +0.65$, while the apparent magnitude it would have if transported to the distance of globular cluster M5 is $V = +19.64$. That would put the Sun in the group of stars about 1.5 magnitudes fainter than the ones currently turning off from the Main Sequence in this cluster.

Evolutionary calculations of double-energy-source HB models by Faulkner and Iben—by then a junior faculty member at MIT—showed that models with a He-burning core and a H-burning shell provide a good fit to the observed horizontal branch of the globular cluster M92 (Faulkner & Iben 1966). When they assumed the mass fraction of CNO elements—which affect the thermonuclear reaction rates in the H-burning shell—to be 10^{-4} and the hydrogen mass fraction in the envelope to be 0.90, they obtained the best fit to the horizontal branch for a model with a 0.6 M_{Sun} He core and a total mass of 0.72 M_{Sun}. This basic idea has remained the foundation of our astrophysical understanding of the horizontal-branch stars ever since.

In 1970, Robert T. Rood (1942–2011), then Iben's student at MIT, constructed a grid of static models for the initial structures of stars on what he termed the "zero-age horizontal branch" or ZAHB (Rood 1970). Each model had a fixed He-core mass M_c, with different models having different total masses, representing different amounts of mass loss between the red-giant branch and the horizontal branch. For example, for a sequence of models—each of which had the same core mass, $M_c = 0.525$ M_{Sun}—the model with the most extensive mass loss had a total mass $M = 0.65$ M_{Sun}, while the model with the least mass loss had a total mass $M = 1.25$ M_{Sun}. The 0.65 M_{Sun} model ended up on the blue end of the horizontal branch, with an effective temperature $T_{eff} = 9984$ K, luminosity $L = 32.5$ L_{Sun}, and radius $R = 1.93$ R_{Sun}. In this model, He was being burned into C and O in a convective core containing about 0.12 M_{Sun} of matter—less than about ¼ the mass of the He core—but it produced about 73% of the total luminosity of the star.

At the other, red, end of the ZAHB, a model with the same 0.525 M_{Sun} He core mass, but with a total mass of 1.25 M_{Sun}—having lost much less mass before reaching the ZAHB—had about the same central temperature and density, but because the overlying mass of H was so much larger, it was much more expanded, with radius $R = 10.37$ R_{Sun}, and accordingly had a much lower effective temperature, $T_{eff} = 5124$ K. The larger H mass also meant that more of the star's luminosity was produced by the H-burning shell—only about 39% came from the He-burning core in this model—and it was more luminous, with $L = 65.5 \, L_{Sun}$.

Calculations of the evolution of horizontal-branch stars (see, e.g., Sweigart 1987) showed that the lifetimes of stars during this phase of core He burning and H-shell burning range from about 110 million years, for a model with a total mass of about 0.5 M_{Sun}, down to about 80 million years, for a model with a total mass of about 0.9 M_{Sun}. In all cases, the mass of the C/O core at the end of horizontal-branch evolution is about 0.15–0.17 M_{Sun}.

13.5 Stellar Mass Loss

Motivated by these theoretical indications of mass loss from evolving stars, German astronomer Dieter Reimers (1943–2021) re-examined the relevant observational data in the early 1970s. These data included blue-shifted absorption lines in some stars (Deutsch 1956), indicating that matter was being ejected from the star into space. Assuming the ejection to be spherically symmetric, observers were able to use the fact that giant, red variable stars known as Mira variables and other stars on what came to be called the "asymptotic giant branch" (AGB; see below) are surrounded by circumstellar dust shells, information about which were just beginning to be obtained from early infrared surveys (Neugebauer & Leighton 1969; Woolf & Ney 1969). Reimers found that the observations could be roughly approximated by a simple formula in which the rate of mass loss is proportional to the luminosity of the star times its radius and is inversely proportional to its total mass (Reimers 1975). Thus, a red giant with a radius perhaps 100 times larger than the Sun's and a luminosity of perhaps 1000 L_{Sun} loses mass at a rate of several times $10^{-8} \, M_{Sun} \, \text{yr}^{-1}$. While this does not seem very large, it means that within the typical ~ 100 million-year timescale of its ascent up the red-giant branch, a low-mass star can lose a significant fraction of its total mass. For example, Thomas Blöcker (Blöcker 1995) found that a 1 M_{Sun} model loses about 30% of its initial mass during its long lifetime on the RGB, while Lee and his Yale colleagues (Lee et al. 1987) obtained good fits to the horizontal branches of the globular clusters M3 and M92 with models that included a 0.5 M_{Sun} He core surrounded by H-rich envelopes ranging in mass from about 0.1 M_{Sun} (at the hot, blue ends of the horizontal branches) up to 0.42 M_{Sun} (at the cool ends). Total masses thus ranged from 0.6 M_{Sun} up to 0.92 M_{Sun}.

13.6 Comparing Theory with Observations

Observations of globular star clusters like that shown in Figure 13.3 have helped both to guide and to test the development of theory. For example, using computer models for the evolution of stars of different masses, astrophysicists have constructed theoretical "isochrones" that can be compared with the observations. A given isochrone shows the locations in the color–magnitude diagram that stars of different masses reach after a given number of years of evolution.

Figure 13.4 shows one example of a comparison between theory and observations for the globular cluster M5 (VandenBerg et al. 2013). This figure shows two different fits to observations obtained with the Advanced Camera for Surveys (ACS) on the

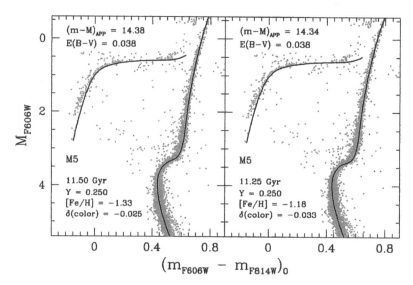

Figure 13.4. Fitting theoretical model isochrones to observations of the globular cluster M5. From Vandenberg et al. (2013); image used with permission from D. Vandenberg.

Hubble Space Telescope (HST). The gray dots represent data for individual stars, while the black curves show the results of stellar evolution calculations for a population of stars with an entire spectrum of masses at a given age; i.e., an "isochrone." The left-hand panel assumes a heavy-element abundance only about 4.7% as large as that in the Sun. For these models, the fit yields a cluster age of 11.50 billion years and a distance of about 7.5 kiloparsecs (kpc). The right-hand panel, which assumes a heavy-element abundance about 6.6% as large as the Sun's, yields an age of 11.25 billion years and again a distance of about 7.5 kpc. The vertical axis is the absolute magnitude determined from the cluster distance and the apparent magnitude measured using the ACS F606W filter (a filter several thousand Ångstroms wide that approximates the bandpass of the conventional visual filter, V). The horizontal axis is the difference in apparent magnitudes obtained using the F606W filter and the F814W filter, which approximates the conventional $V-I$ (visual minus infrared) color index.

13.7 Beyond the Horizontal Branch

Beginning in the early 1970s, groups of astrophysicists from around the globe undertook calculations of the evolution of stars following the horizontal-branch (HB) and asymptotic giant branch (AGB) phases. The AGB is a region of the H–R diagram roughly parallel to but at higher luminosity than the red-giant branch (RGB) phase of stellar evolution; I discuss it in more detail in the following chapter. However, not all HB stars follow a high-luminosity post-AGB (P-AGB) path.

If the mass of the H envelope is too small, the star is unable to pass through the AGB phase, as shown in Figure 13.5 from Dorman et al. (1993). Such stars populate the extreme horizontal branch (EHB). The EHB models have lifetimes of 120–150

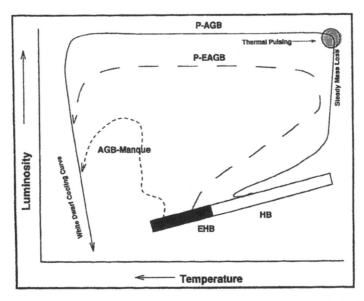

Figure 13.5. Schematic H–R diagram showing post-horizontal-branch evolutionary tracks for low-mass stars. From Dorman et al. (1993); image used with permission from R. W. O'Connell.

million years, and the bluest HB and EHB stars correspond to hot subdwarf stars (with spectral types sdB, sdOB, and sdO, indicating that they are very hot, like Main Sequence O and B stars). If the envelope mass is slightly less than 0.05 M_{Sun}, the star begins to climb the AGB, but the H-burning shell burns out before it reaches the thermally pulsing AGB stage, and the star passes through an abbreviated form of evolution to very high effective temperatures (called "post-early AGB," or P-EAGB, evolution) before turning down toward the white-dwarf domain. At still lower envelope masses, the star instead passes through a brief excursion to higher luminosities and effective temperatures—called "AGB-manqué evolution"—before passing through the region of the H–R diagram populated by sdB and sdOB stars (Greenstein & Sargent 1974) *en route* to becoming a white dwarf. The P-EAGB lifetimes are a few times 0.1 Myr, and the AGB-manqué phase lasts 20–40 Myr.

References

Allen, C. W. 1955, Astrophysical Quantities (2nd ed.; Univ. London: Athlone Press) 266

Blöcker, T. 1995, A&A, 297, 727

Cook, C. W., Fowler, W. A., Lauritsen, C. C., & Lauritsen, T. 1957, PhRv, 107, 508
   This story is told in detail in Halpern, P. 2021, Flashes of Creation (New York: Basic Books) 152

Dearborn, D. S. P., Lattanzio, J. C., & Eggleton, P. P. 2006, ApJ, 639 405; see also
   Mocák, M., et al. 2009, A&A, 501, 659

Demarque, P., & Mengel, J. G. 1971, ApJ, 164, 317

Deutsch, A. J. 1956, ApJ, 123, 210

Dorman, B., Rood, R. T., & O'Connell, R. W. 1993, ApJ, 419, 596

Faulkner, J. 1966, ApJ, 144, 978

Faulkner, J., & Iben, I. Jr. 1966, ApJ, 144, 995

Greenstein, J. L., & Sargent, A. I. 1974, ApJS, 28, 157

Hoyle, F., & Schwarzschild, M. 1955, ApJS, 2, 1

Iben, I. Jr. 1967, ApJ, 147, 624

Lee, Y.-W., Demarque, P., & Zinn, R. 1987, IAU Coll. #95, Second Conf. on Faint Blue Stars, ed. A. G. D. Philip, D. Hayes, & J. Liebert (Schenectady, NY: L. Davis Press) 137

Neugebauer, G., & Leighton, R. B. 1969, Two-Micron Sky Survey, A Preliminary Catalog (NASA SP-3047)

Reimers, D. 1975, in Problems Stellar Atmospheres and Envelopes, ed. B. Baschek, W. H. Kegel, & G. Traving (New York: Springer) 229

Rood, R. T. 1970, ApJ, 161, 145

Salpeter, E. E. 1952, ApJ, 115, 326

Schwarzschild, M., & Härm, R. 1962, ApJ, 136, 158

Schwarzschild, M., & Selberg, H. 1962, ApJ, 136, 150

Sweigart, A. V. 1987, ApJS, 65, 95

VandenBerg, D. A., Brogard, K., Leaman, R., & Casagrande, L. 2013, ApJ, 775, 134

Woolf, N. J., & Ney, E. P. 1969, ApJ, 155, L181

Chapter 14

How an Intermediate-Mass Star Becomes a White Dwarf

An intermediate-mass star is one with an initial Main Sequence mass between approximately 2.3 M_{Sun} and an upper limit of about 6–8 M_{Sun}. As discussed in the previous chapter, stars with initial masses less than 2.3 M_{Sun} develop partially degenerate He cores on their way up the red giant branch after leaving the Main Sequence. They consequently ignite He burning in a relatively violent "He flash" that causes varying amounts of mass loss before settling down to a prolonged phase of stable core He burning on the horizontal branch. In contrast, the He core of an intermediate-mass star does not become degenerate, so it ignites core He burning more quietly. The upper limit to the mass range of intermediate-mass stars is set by the condition that the C/O core produced by He burning never becomes hot or dense enough to ignite C-burning nuclear reactions before it becomes degenerate and ultimately evolves into a white dwarf.

14.1 From the ZAMS to the AGB

As noted in Chapter 7, in 1959 Louis Henyey and his colleagues developed a method for using digital computers to calculate evolutionary sequences of stellar models automatically (Henyey et al. 1959). Another of the early leaders to use electronic digital computers to study the detailed evolutionary changes in the internal structures of stars was Icko Iben Jr. (Figure 14.1). Then a postdoc based at Caltech's Kellogg Radiation Laboratory, Iben was particularly interested in the details of nuclear reactions in stars. Beginning in 1964, he published calculations of evolutionary tracks in the Hertzsprung–Russell diagram for stars ranging from 1 to 15 M_{Sun} (see Figure 14.2) (Iben 1964). For the 1.0 and 1.5 M_{Sun} models, he terminated his calculations at the beginning of the He flash. Because stars more massive than about 2 M_{Sun} do not experience a He flash but instead ignite He burning quasi-statically—in much the same way that H-burning reactions are

Figure 14.1. Icko Iben Jr. Credit: AIP Emilio Segrè Visual Archives, John Irwin Slide Collection; copyright American Institute of Physics.

ignited as a star first arrives on the Main Sequence—Iben was accordingly able to follow their evolution through the stage of core He burning and beyond. In subsequent publications, he studied in detail the changes in structure and the internal distribution of the chemical composition for several model sequences, each with fixed total mass (Iben 1965).

At about the same time, Rudolf Kippenhahn (1926–2020; Figure 14.3) and his colleagues at the Max Planck Institute in Munich were performing similar but independent computations. Figure 14.4 shows the post-Main-Sequence evolutionary track of a 5 M_{Sun} star from their computations of some 400-odd evolutionary models in the Hertzsprung–Russell diagram (Kippenhahn et al. 1965). The general shape of their evolutionary track compares very well with that of Iben's 5 M_{Sun} model shown in Figure 14.2, although details are somewhat different in the portions of the two tracks after the models pass the point of helium ignition.

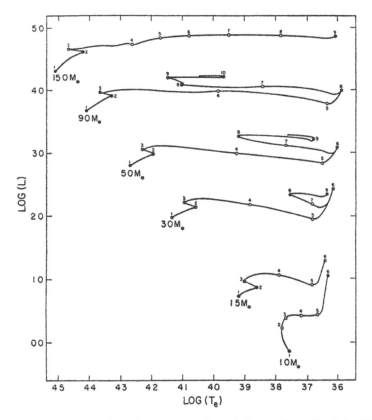

Figure 14.2. Iben's evolutionary tracks in the Hertzsprung–Russell diagram. From Iben, I. Jr. 1964, ApJ, **140**, 1631; reproduced with permission from I. Iben Jr.

Several points along the evolutionary track in Figure 14.4 are marked by the letters A through K. The same letters are used to identify the same points in the evolution of the internal structure of this 5 M_{Sun} model shown in Figure 14.5. In this figure—often called a "Kippenhahn diagram"—a vertical line represents the structure of the stellar model in concentric spherical shells, each labelled by the fraction of the stellar mass inside it, as listed on the vertical axis, at the particular model age listed on the horizontal axis. For example, the point marked "0.2" on the vertical axis corresponds to a shell containing 20% of the total stellar mass. Note that this figure only illustrates the inner 60% of the star's mass—it does not extend all the way to the stellar surface—in order to make clearer the evolutionary changes in the deep interior. The horizontal axis represents the elapsed time since the onset of H burning on the Zero-Age Main Sequence (ZAMS), measured in tens of millions of years (units of 10^7 years). There are two breaks in this scale, one shortly before the ignition of He burning and a second that expands the scale to make it easier to follow the inward growth of the surface convection zone as it approaches the outward-moving He-burning shell in these red-giant models.

Figure 14.3. Rudolf Kippenhahn in 1985. Credit: AIP Emilio Segrè Visual Archives, John Irwin Slide Collection; copyright American Institute of Physics.

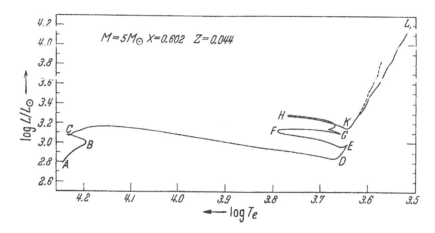

Figure 14.4. Kippenhahn 5 M_{Sun} evolutionary track in the H–R diagram. From Kippenhahn et al. (1965); reproduced with permission from Springer Nature. See text for explanations of the letters A–K.

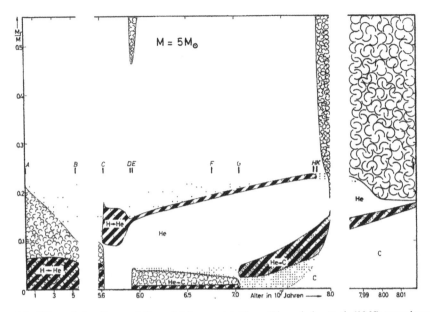

Figure 14.5. "Kippenhahn diagram" for a 5 M_{Sun} model. From Kippenhahn et al. (1965) reproduced with permission from Springer Nature. The vertical axis gives the fraction M_r/M of the total mass M that lies interior to radius r, and the horizontal axis gives the time since H ignition on the ZAMS in tens of millions of years. The letters A–K correspond to the points marked with the same letters on the evolutionary track in Figure 14.4.

Kippenhahn and his colleagues adopted a few artistic touches to make it easier to see what is going on deep in the interior of the star. One was to render all convection zones as "cloudy" regions.[1] A second was to represent nuclear-burning regions as boldly cross-hatched areas in the diagram. Regions of mixed composition created by the interplay between convection and nuclear burning were represented as stippled areas.

With these explanations, Figures 14.4 and 14.5 provide a detailed summary of the immediate post-Main-Sequence evolution of a 5 M_{Sun} star. At point A, when the star is located on the ZAMS, the H-burning core occupies about the innermost six percent of the star's mass. Because of the high rate of energy generation by the H-burning CNO reactions at this stellar mass, the temperature gradient in the central regions of the star is steep enough to cause the innermost 21% of the mass to become convective. As the star moves from point A to point B over the next 56 million years, the hydrogen fuel throughout the convection zone is gradually depleted.[2]

---

[1] In a private conversation decades later, Alfred ("Arlie") Weigert (1927–1992)—one of Kippenhahn's close collaborators—told the author that in every such diagram, the artist had embedded a caricature of a human face. It remains as an exercise for the reader to locate that image!

[2] Convection carries fresh fuel into the burning zone while simultaneously dispersing the helium "ashes" uniformly throughout the convection zone.

The convection zone gradually shrinks in mass, declining from 21% to about nine percent of the stellar mass. It leaves behind a region of mixed composition, with hydrogen and helium abundances intermediate between the hydrogen fraction in the outer envelope and that in the shrinking H-depleted convective core. This region also contains enhanced levels of ^{14}N and reduced levels of ^{12}C, which have been modified from their original values during core H burning. At point B, the residual hydrogen in the convective core has become totally depleted. In a very short time, nuclear energy production in the inner—now He dominated—core ceases completely. The core then begins to contract, releasing gravitational energy to replace the loss of thermonuclear energy production in the core, while H burning "switches on" in a relatively thick shell. This shell source actually generates energy at a more rapid rate than had been the case during the period of core hydrogen burning, so the luminosity escaping from the surface of the star increases, and its surface begins to expand.

Over about the next three million years (from point C to point D), the helium core continues to contract and heat, while the H-burning shell gradually burns its way outward through the still-H-rich envelope. At point D the core becomes hot enough (about 125 million Kelvin) to ignite the He-burning triple-alpha reactions, and a new convection zone develops at the center of the star. Between points D and E, the star climbs the (short) red-giant branch (RGB), the H-burning shell becomes narrower, the convective He-burning core stabilizes, and a deep convection zone reaches downward from the stellar surface to a mass fraction of about 0.46. That is, the outer 54% of the mass of the star is contained in this single, very deep convection zone. Over about the next 11 million years (point E to point G), the star executes a "blue loop" (which is what forms the "horizontal branch" for low-mass stars), with the innermost convection zone again declining gradually while the thin H-burning shell continues to eat its way outward through the H-rich envelope.

At point G, the star again undergoes a rapid transition from core nuclear burning to shell burning. At this point, the star has two nuclear burning shells. The outer H-burning shell continues to eat its way gradually outward through the deep H-rich envelope, which at this point still occupies about the outermost 80% of the stellar mass. It leaves in its wake a thick layer of helium, initially extending over about 15% of the stellar mass. At the base of the He layer, a thick He-burning shell continues to process the helium fuel into a mixture of carbon and oxygen. At point K, about 79 million years after it has left the ZAMS, the star begins to develop a very deep outer convection zone, extending all the way down from the stellar surface to the location of the He-burning shell. In the Hertzsprung–Russell diagram, the star continues along the track it followed during the RGB phase, but at higher luminosity. This phase is accordingly termed the "asymptotic giant branch" (AGB), a term which seems first to have been used in a review article by Iben (Iben 1974). This takes only about 1.3 million years. By the end of the calculation, the model is left with a carbon–oxygen (C/O) core containing about 17% of the stellar mass. This is about $0.85\ M_{Sun}$ for a 5 M_{Sun} star, within the expected mass range for a white dwarf—if only the star could rid itself of its huge, hydrogen-dominated envelope!

Kippenhahn and his colleagues terminated their calculation at this point for several reasons. First, they had not included the nuclear reactions responsible for carbon burning. Second, they had not included mass loss, which becomes increasingly important at higher luminosities. Third, contact between the outermost convection zone and the He-burning shell would have triggered a violent outburst, as fresh hydrogen would have been brought into a region where temperatures are already high enough for helium burning. To follow the evolution further would require detailed hydrodynamic calculations, which were not possible at the time.

As such calculations show, the progress of stellar evolution can be thought of as a series of thermonuclear-burning episodes, with gravitational contraction of the "ashes" of each burning stage heating them up until the central temperature becomes high enough to ignite the next stage of nuclear burning. In the process, a star with an initially uniform composition dominated by hydrogen develops a layered "onion-skin" structure, with H-rich material overlying a He layer, which overlies a C/O layer, and so on. At the same time, the luminosity of the star increases more or less steadily to higher and higher luminosities, while the star's center becomes steadily denser and hotter.

As Figures 14.2 and 14.4 show, as an intermediate-mass star transitions from core H burning, to shell H burning, to core He burning, to thermonuclear burning in both H and He shell sources, it executes several transitions back and forth in the H–R diagram between radii of several solar radii (depending upon the stellar mass) alternately swelling to many hundreds of solar radii. In the latter phases, the stellar surface cools to temperatures of about 3000 K to 4000 K, and the star develops a subsurface convection zone that extends so deep that it may engulf regions of mixed chemical composition left behind as the convection zone driven by nuclear burning retreats. During these episodes—appropriately termed "dredge-up" events—the subsurface convection zone mixes the composition changes all the way to the stellar surface, where they become visible to terrestrial observers.

Up to three dredge-up events can occur after an intermediate-mass star finishes core H burning, but only stars with sufficiently large H-exhausted cores experience all of them (Becker & Iben 1979).

The first dredge-up event occurs as the stellar envelope expands during the first ascent of the red-giant branch at the end of core H burning. For the 5 M_{Sun} model shown in Figure 14.5, the base of the subsurface convection zone at point E in the star's evolution does not reach quite deep enough for this dredge-up phase to occur. This may be the consequence of too small a core mass in this particular intermediate-mass model, or it may be the result of differences in the assumed physical properties of the stellar matter. In any event, when the first dredge-up phase does occur, it primarily enhances the surface abundance of ^{14}N and depletes that of ^{12}C, both of which result from nucleosynthetic changes during core H burning.

The second dredge-up phase occurs as an intermediate-mass star ascends the asymptotic giant branch (AGB). At this point—beyond point K in Figure 14.5—the subsurface convection zone reaches so deep into the star that it engulfs much of the region of mixed composition left behind by the now-inactive H-burning shell. Because the mixed composition in this region also is produced during H burning—

this time in the thermonuclear shell source rather than in the central core—the composition changes that appear at the stellar surface mimic those produced during the first dredge-up phase; i.e., the ^{14}N abundance is enriched at the surface.

The third dredge-up phase occurs after full-amplitude thermal pulses occur, which we discuss below. These subsequent phases occur beyond the stages shown in Figures 14.4 and 14.5.

14.2 The Thermally Pulsing AGB

During the early AGB phase, a star derives most of its energy from the He-burning shell that surrounds the C/O core. As the core contracts, releasing gravitational energy and becoming denser and hotter, the outer layers swell up to hundreds of times larger than the solar radius. Meanwhile, the He-burning shell advances closer and closer to the H/He discontinuity left behind by the now-extinguished H-burning shell. However, heating of the H-rich layers reignites the H-burning shell, and it soon becomes the main energy source for the star. The star now has two nuclear-burning shells, with the upper H-burning shell feeding He into the layer just below it, and the He-burning shell producing more C and O, which are added to the growing and increasingly degenerate core.

However, as Arlie Weigert discovered in 1966, the He-burning shell is unstable, and its energy output begins to oscillate (Weigert 1966; see Figures 14.6(a) and (b)). Because of the very strong temperature sensitivity of the He-burning 3α reaction, a slight increase in the temperature of the He-burning shell causes a significant increase in its rate of energy generation, which causes a further increase in the temperature, leading to a thermonuclear runaway. Of course, the additional energy release increases the stellar luminosity and expands the stellar radius, which then reduces the temperature of the shell, enabling He burning to return to normal again. Figure 14.6a shows a succession of such thermal pulses marching steadily outward through the mass of the star, while the C/O core continues to grow. Figure 14.6b shows an expanded view of the first two thermal pulses, illustrating the brief time duration of the He-burning pulse (less than 100 years) and of the small convection zone formed there to carry the excess heat away and allow the He-burning shell to return to normal. Weigert calculated the first six of these thermal pulses, which extend over about 19,000 years, each separated from the next by about 3000 years.

About a decade later, Iben calculated 10 such thermal pulses for a 7 M_{Sun} star, which he found to recur at intervals of roughly 2500 years (Iben 1975). He also found that the nuclear reactions that occur during the thermal pulses serve as a major source of neutrons for stellar nucleosynthesis, as discussed further in Chapter 17. In particular, he found that—between the thermal pulses of the He-burning shell —the H-burning shell leaves behind it an excess of ^{14}N as it burns its way outward through the star. During a subsequent thermal pulse, the He-burning shell engulfs the excess nitrogen and, after two α captures and a β decay, converts it to ^{22}Ne. Near the peak of the next thermal pulse, the reaction ^{22}Ne (α, n) ^{25}Mg produces a strong flux of slow neutrons, which play an important role in the production of many of the chemical elements found in nature.

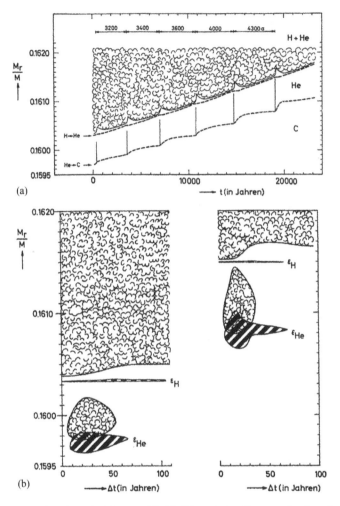

Figure 14.6. Thermal pulses in a 5 M_{Sun} model. (a) Kippenhahn diagram showing six thermal pulses. (b) Close-up of the first two thermal pulses showing the H-burning shell (labelled ε_H) and the He-burning shell (labelled ε_{He}) together with the associated convection zones (see text). From Weigert (1966); reproduced with permission from Springer Nature.

14.3 Mass Loss, Neutrino Emission, and the Central Stars of Planetary Nebulae

After some number of "thermal pulses" (or nuclear-burning "shell flashes"), an AGB star undergoes a period of very rapid mass loss (called a "superwind") that results in the ejection of a substantial amount of mass. When the ejected gas has expanded sufficiently, irradiation by the strong UV radiation from the now-exposed, very hot C/O core of the star forms a so-called "planetary nebula." The Ring Nebula in the constellation Lyra, shown in Figure 14.7, is one example of such a planetary nebula. The small bright dot at the center of the strongly ionized nebula is the central

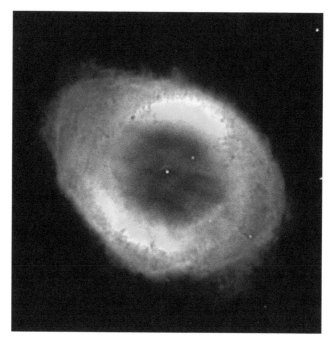

Figure 14.7. The Ring Nebula, a planetary nebula in the constellation Lyra. Credit: C. F. Claver, WIYN, NOIRLab/NSF.

star. With a mass of about 0.6 M_{Sun}, it has an effective temperature of about 125,000 K but a radius only about 0.03 R_{Sun}. It is still more than twice the size of a white dwarf of the same mass. The reason for the misleading name "planetary nebula" is that such nebulae have visible disks, so they resembled planets in the telescopes available in the late 1700s when they were first described and named. During this post-AGB (P-AGB) phase of its evolution, the core of the star evolves rapidly at nearly constant luminosity to very high effective temperatures (see Figure 14.8). The stellar radius decreases rapidly during this stage, ultimately making the star's density so high that electron degeneracy halts further contraction, and the star begins to cool at almost constant radius to become a white dwarf (see Chapter 15).

However, the stages in the evolution of a star that take it from the AGB through the planetary nebula phase remain among the least-well understood in the lifetime of an intermediate-mass star, as emphasized as recently as 2018 by Marcello Miller Bertolami (Miller Bertolami 2018). In part, this is because of uncertainties in our knowledge of stellar winds, which strongly affect these phases. In part also the uncertainty is due to decoupling between the outer layers of the star, which are being expelled into space during these phases, and the innermost parts of the star, which are undergoing contraction down to white-dwarf dimensions while hidden from an observer's view beneath those expanding outer layers. A further complication is that the transition from the end of the AGB to the point where the hot, dense core of the star again becomes visible through the thinning veil of expelled matter—which forms a planetary nebula around the central star—occurs very rapidly on the

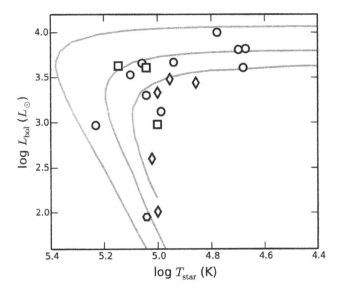

Figure 14.8. Evolution through the planetary nebula phase. From Montez et al. (2015). The orange curves represent the post-AGB tracks calculated by Schönberner (1983) and by Blöcker (1995), and the open symbols represent the positions of CSPNe in this form of the H–R diagram. The circles represent point-like X-ray emitting sources detected in the Planetary Nebula Survey with the Chandra X-ray Observatory, diamonds correspond to known binary CSPNe, and squares represent spectral types corresponding to extremely hot stars.

timescale of stellar evolution. It may take only 10,000 years or less for a star to make this transition, as compared to the millions or billions of years for steady-state nuclear-burning.

One of the factors that speeds the evolution of a star through the planetary-nebula phase is the emission of neutrinos from the stellar core. We saw in Chapter 8 that some of the nuclear reactions that occur in the center of the Sun emit neutrinos that escape almost unimpeded because they interact so weakly with matter; that is, stellar matter is almost completely transparent to neutrinos. Beginning in the 1960s, astrophysicists realized that at the high temperatures and densities that occur in later stages of stellar evolution several other kinds of reactions also can produce neutrinos—or, more accurately, neutrino–antineutrino pairs—that can drain energy from a star's interior. Among others, these include so-called plasma neutrinos, in which quantized plasma waves decay into neutrino–antineutrino pairs (Adams et al. 1963). As a post-AGB star contracts toward the region of the H–R diagram populated by the central stars of planetary nebulae (CSPNe), its core becomes hotter and denser, increasing the rates at which neutrinos are produced. The resulting neutrino luminosity drains energy from the central regions of the star, accelerating its evolution toward and through the CSPNe and on toward the fully degenerate white-dwarf state.

The difficulties noted above have not prevented astronomers and astrophysicists from tackling the question of the relationship between the CSPNe (also called planetary nebula nuclei or PNNe) and earlier and later phases of stellar evolution. The well-known Soviet astronomer Iosif S. Shklovsky (1916–1985) first suggested that the CSPNe are the dense cores of stars that have lost most or all of their hydrogen-rich envelopes (Shklovsky 1957). In 1970, Bohdan Paczyński (1940–2007) made the first attempt to connect such stars into the framework of stellar evolution (Paczynski 1970). He evolved several models, with masses between 0.8 M_{Sun} and 15 M_{Sun}, from the ZAMS through the stages at which H and He become exhausted in the central regions. Assuming that models with initial masses of 3 M_{Sun} or less lose their H-rich envelopes during the double-shell-burning phase, he evolved the residual cores—consisting primarily of some mixture of C and O, but with thin residual envelopes having just enough mass to sustain the He and H shell-burning sources—to the white dwarf stage. He included neutrino energy losses, but—other than the removal by hand, as it were, of most of the H/He envelope—these models did not take mass loss into account.

Following the recognition in the mid-1970s of the importance of stellar mass loss, theorists began to include approximations for this effect in stellar evolution calculations. This is particularly important for AGB and post-AGB stars, because their higher luminosities and larger radii make it easier for a star to lose mass: the higher luminosity produces a higher radiation pressure to drive the mass loss, and the larger radius results in a greatly reduced surface gravity, which requires less force to drive increased mass loss from the star. In fact, as Eddington discovered in the 1920s, the luminosity of a star cannot exceed the value where the radiation pressure alone is sufficient to balance the force of gravity at the stellar surface (Eddington 1926). An approximate value for this "Eddington limit" is about 30,000 L_{Sun}, and the value scales up with the stellar mass. The mass-loss rate thus increases greatly at these higher luminosities and larger stellar radii—as indicated, for example, by Reimers' empirical expression discussed in the previous chapter—and both the luminosity and radius increase dramatically during thermal pulses. By the late 1980s and early 1990s, it had become clear that at least some planetary nebulae contain multiple shells, indicating that gas was ejected from the central star several times (Chu et al. 1987). This suggests that the thermal pulses may indeed be responsible for driving at least some of the mass loss (Frank et al. 1994).

The first stellar-evolution calculations to include mass loss during the AGB phase were carried out by Detlev Schönberner in 1979 for two low-mass models (Schönberner 1979). Starting from the beginning of central He burning, his 1 M_{Sun} model ultimately became a 0.598 M_{Sun} white dwarf after numerous thermal pulses, while his 1.45 M_{Sun} model ultimately became a 0.644 M_{Sun} white dwarf. He found that at the tip of the AGB, the rate of mass loss competes with H burning in reducing the mass of the residual hydrogen envelope. At the end of the thermal pulses, each model underwent rapid gravitational contraction to high effective temperatures, until the C/O core became degenerate, slowing and ultimately stopping the contraction, and the star began to cool down to become a white dwarf. A decade and a half later, E. Vassiliadis and Peter Wood computed evolutionary sequences for models with initial Main Sequence masses ranging from 0.89 M_{Sun} up to 5 M_{Sun} through the end of the AGB phase, including improved approximations for the rate of mass loss (Vassiliadis & Wood 1993). They found that a very rapid mass-loss process—dubbed a "superwind" (Iben & Renzini 1983)—develops during the last few thermal pulses, when the stellar luminosity is very high. A few years later, Thomas Blöcker performed an independent set of calculations for models with initial Main Sequence masses ranging from 1 to 7 M_{Sun} (Blöcker 1995). He employed different assumptions for the rate of mass loss, but he found results quite similar to those obtained by others. In 2015, Rodolfo Montez and his collaborators employed X-ray observations to refine the positions of CSPNe in the Hertzsprung–Russell diagram, and they found good agreement with the evolutionary tracks calculated by Schönberner and by Blöcker (see Figure 14.8; Montez et al. 2015).

More recently, Marcelo Miller Bertolami published a new grid of 27 evolutionary sequences with initial masses ranging from 0.8 to 4 M_{Sun} and four different metallicities (Miller Bertolami 2016). This grid of post-AGB sequences takes into account the substantial developments in stellar astrophysics since the mid-1990s. These newer models reproduce several of the observed properties of AGB and post-AGB stars that are not reproduced by the older grids, and they also do a better job of predicting the properties of the central stars of planetary nebulae. The new models also have post-AGB timescales that are substantially shorter than the older models. The final masses of these models range from about 0.5 to about 0.85 M_{Sun}.

Once the C/O core of a post-AGB star becomes degenerate, preventing further significant contraction, any residual H- or He-burning shells eventually die out, and the only energy source left to sustain the stellar luminosity is the residual heat content of the very hot but strongly degenerate core. We discuss this further in the following chapter. As discussed above, the evolutionary paths of such stars fit the observed properties of the central stars of planetary nebulae, which appear to be well on their way to becoming white dwarfs.

References

Adams, J. B., Ruderman, M. A., & Woo, C.-H. 1963, PhRv, 129 1383; see also
    Inman, C. L., & Ruderman, M. A. 1964, ApJ, 140, 1025
    Beaudet, G., Petrosian, V., & Salpeter, E. E. 1967, ApJ, 150, 979
Becker, S. A., & Iben, I. Jr. 1979, ApJ, 232, 831
Blöcker, T. 1995, A&A, 297, 727
Chu, Y.-H., Jacoby, G. H., & Arendt, R. 1987, ApJS, 64 529; see also
    Balick, B., Gonzalez, G., Frank, A., & Jacoby, G. 1992, ApJ, 392, 582
Eddington, A. S. 1926, The Internal Constitution of the Stars (Cambridge: Cambridge Univ. Press), 16
Frank, A., van der Veen, W. E. C. J., & Balick, B. 1994, A&A, 282, 554
Henyey, L. G., Wilets, L., Böhm, K. H., LeLevier, R., & Levée, R. D. 1959, ApJ, 129, 628
Iben, I. Jr. 1964, ApJ, 140, 1631
Iben, I. Jr. 1965, ApJ, 141 993; see also
    Iben, I. Jr. 1965, ApJ, 142, 1447
    Iben, I. Jr. 1966, ApJ, 143, 483
    Iben, I. Jr. 1966, ApJ, 143, 505

Iben, I. Jr. 1966, ApJ, 143, 516

Iben, I. Jr. 1967, ApJ, 147, 624

Iben, I. Jr. 1967, ApJ, 147, 650

Iben, I. Jr. 1974, ARA&A, 12, 215

Iben, I. Jr. 1975, ApJ, 196, 525

Iben, I. Jr., & Renzini, A. 1983, ARA&A, 21, 271

Kippenhahn, R., Thomas, H.-C., & Weigert, A. 1965, ZA, 61, 241

Miller Bertolami, M. M. 2016, A&A, 588, A25

Miller Bertolami, M.M. 2018, IAU Symp. #343, Why Galaxies Care About AGB Stars: A Continuing Challenge through Cosmic Time, ed. F. Kerschbaum, M. Groenewegen, & H. Olofsson (Cambridge: Cambridge Univ. Press), 36

Montez, R. Jr., et al. 2015, ApJ, 800, 8

Paczynski, B. 1970, AcA, 20, 47

Schönberner, D. 1979, A&A, 79 108; see also
Schönberner, D. 1983, ApJ, 272, 708

Shklovsky, I. 1957, IAU Symp. 3: Non-Stable Stars, ed. G. H. Herbig (Cambridge: Cambridge Univ. Press) 83

Vassiliadis, E., & Wood, P. R. 1993, ApJ, 413, 641

Weigert, A. 1966, ZA, 64, 395

Chapter 15

White Dwarfs: Fading Embers of Burnt-Out Stars

Before the 20th century, no one had ever imagined that stars as strange as the ones we now call "white dwarfs" might even exist. The first clue came in 1783, though no one realized it at the time. That year, British astronomer William Herschel discovered that the Main Sequence star 40 Eridani A, which is somewhat cooler than the Sun, had a pair of faint companion stars, the close binary system 40 Eri B and C. At the time, this aroused only moderate interest among astronomers, for binary stars were then being identified in droves, and most stars were known to be faint, red objects orders of magnitude dimmer than the Sun. This situation persisted for about a century and a half, but it changed abruptly in 1910.

During the early years of the 20th century, Edward C. Pickering (1846–1919), the Director of the Harvard College Observatory, had recruited a group of talented young women[1] to carry out the painstaking labor required to analyze and classify hundreds of thousands of stellar spectra that had been recorded on a collection of photographic glass plates by amateur astronomer Henry Draper. After his death, Draper's widow donated the collection to the Observatory. In 1910, Pickering was attending a meeting with Princeton astronomer Henry Norris Russell (1877–1957), who had been charting the brightnesses of stars against their spectral types. Knowing of the work at Harvard, Russell asked Pickering if his staff had determined the spectral type of 40 Eri B. Pickering telephoned Mrs. Williamina Fleming (1857–1911; see Figure 15.1) at the Observatory to inquire about it. When she called back later that day, she reported that 40 Eri B was of spectral type A, meaning that it displayed strong absorption lines of hydrogen, corresponding to a surface temperature of about 10,000 K. She thus became the first person to discover a white dwarf.

---

[1] The story of these "Harvard Computers" is engagingly told by Dava Sobel in her book *The Glass Universe: How the Ladies of the Harvard Observatory took the Measure of the Stars* (Viking 2016).

doi:10.1088/2514-3433/acce33ch15

Figure 15.1. Williamina Paton Stevens Fleming. Source: https://en.wikipedia.org/wiki/Williamina_Fleming.

Suddenly, objects like 40 Eri B became intensely interesting to astronomers. How could a star about twice as hot as its primary be hundreds of time fainter? The obvious answer was that the radius of 40 Eri B had to be almost 100 times smaller than that of 40 Eri A. Because of its white-hot temperature and puzzlingly small dimensions, astronomers named the few peculiar stars like 40 Eri B "white dwarfs." But what could account for such small sizes? Astronomers remained mystified.

The mystery deepened when careful observations of the orbit of Sirius B, the faint white-dwarf companion to the bright Main Sequence star Sirius, revealed that the mass of Sirius B is about 1 M_{Sun}. With a mass this large and a radius about the size of Earth, the average density of Sirius B is about a million times greater than that of the Sun! The astronomers of the day regarded this conclusion as absurd (Eddington 1926).

It took several decades more—and two revolutions in the science of physics!—before the mystery of Sirius B could be solved.

The first scientific revolution was Albert Einstein's publication of the special theory of relativity in 1905. In that seminal paper, Einstein showed that the relationship between the kinetic energy of a body and its velocity changes substantially as the velocity approaches the speed of light. He also discovered the surprising equivalence between mass and energy, encapsulated in his famous equation $E = mc^2$, where E is energy, m denotes the mass of the body, and c is the speed of light.

The second revolution was the development of quantum physics. During the early decades of the 20th century, physicists all over the world were struggling to understand the workings of nature on an atomic scale. The German physicist Max Planck started it all in 1900 by postulating that light can only be emitted or absorbed by an atom in discrete quanta, which we now call "photons." In 1913, Danish physicist Niels Bohr took the quantum theory a step further, showing that a model of the hydrogen atom in which an electron exists only in discrete quantum energy levels provides a natural explanation for the emission and absorption lines in the hydrogen spectrum. And in 1925, to help explain the structure of more complex atoms, Wolfgang Pauli (1900–1958) enunciated the exclusion principle: two or more electrons cannot occupy the same quantum state.

The following year, the Italian physicist Enrico Fermi (1901–1954) and his British colleague Paul A. M. Dirac (1902–1984) published a statistical theory for particles like electrons that obey the Pauli Exclusion Principle. Because no two electrons can exist in the same microscopic state, then even at the absolute zero of temperature they fill up all the lowest available energy states to a level now termed the "Fermi energy;" in this condition, they are said to be "degenerate" (see Appendix D). And because the electrons have so much energy, they exert a finite (non-zero) pressure even at zero temperature.

The young Indian astrophysics student Subrahmanyan Chandrasekhar (1910–1995; see Figure 15.2) learned about Fermi–Dirac statistics during a 1928 visit to Madras by the German physicist Arnold Sommerfeld (1868–1951). Two years later, Chandrasekhar was on a steamer traveling from Bombay to England to begin his graduate studies with Ralph H. Fowler (1889–1944) at Cambridge University. In 1925, Fowler had found that a natural explanation for a stellar mass crammed into a volume about the size of a terrestrial planet is provided if the matter of which the star is composed is so dense that the electrons have become degenerate. With time on his hands during the long sea voyage, Chandra —as his friends subsequently nicknamed him—pondered the mystery posed by the density of Sirius B, and he realized that if the degenerate electrons are also extremely relativistic—as is the case for sufficiently massive white dwarfs—the pressure is no longer capable of supporting the star's mass. Electron degeneracy in fact leads to a peculiar relationship between the mass and radius of a white dwarf (Figure 15.3) in which the more massive white dwarfs have smaller radii (Chandrasekhar 1935). Indeed, there is a limiting mass, above which a stable white dwarf cannot exist. Subsequent calculations showed the limiting mass to be about 1.44 M_{Sun}, which is now known as the "Chandrasekhar limiting mass."

By the early 1920s, only three white dwarf stars were known: Sirius B, 40 Eridani B, and van Maanen 2. Because of the growing interest in the properties of these dense stars, the Dutch astronomer Willem J. Luyten (1899–1994) decided to mount

Figure 15.2. The young Subrahmanyan Chandrasekhar in 1933 after receiving his Ph.D. from Trinity College, Cambridge. Credit: AIP Emilio Segrè Visual Archives, gift of Kameshwar Wali. Copyright American Institute of Physics.

a search dedicated to finding more of them, initiating the Bruce Proper Motion Survey for this purpose in 1929. The "proper motion" of a star is its apparent motion across the sky due to its intrinsic velocity. Luyten's idea was that faint stars like white dwarfs can be seen most easily when they are relatively close to the Earth, and stars with large proper motions are likely to be astronomically nearby. If in addition they are relatively blue in color, then they are likely to be hot; and faint, hot stars are likely to be white dwarfs. By the end of his long career in 1967, Luyten had screened more than 100 million stars, determined the proper motions for more than 400,000, and had identified several thousand white-dwarf candidates. To confirm that a star actually is a white dwarf, however, requires a careful examination of its spectrum. The Dutch–American astronomer Gerard P. Kuiper (1905–1973) made a preliminary effort to understand the spectra of white dwarfs (Kuiper 1941), but real progress had to await the advent of the 200-inch (5-meter) telescope in 1949.

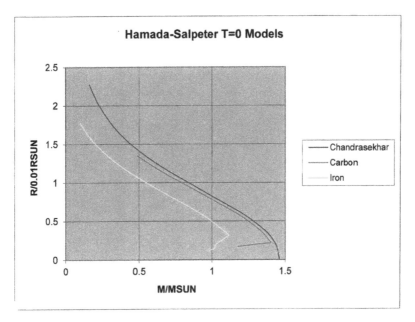

Figure 15.3. The white-dwarf mass–radius relationship as plotted by the author. The blue curve represents Chandrasekhar's white-dwarf models, while the two colored curves represent models computed by Hamada and Salpeter (1961), which include the effects of Coulomb interactions among the ions and electrons.

While Luyten and others were working to find more white dwarf stars, these still mysterious objects were providing theorists with yet another puzzle. After the success of electron degeneracy in explaining the mechanical structures of white dwarfs, theorists sought to calculate their internal temperatures as well. The first calculations, using the theory of radiative transfer that works so well for ordinary stars, led to predicted central temperatures of some billions of degrees Kelvin. At such high temperatures, however, the electrons do not remain degenerate, so this model was not self-consistent. Something was definitely wrong!

The answer was provided by a young American physicist. Robert E. Marshak (1916–1992) was then a graduate student working with Hans Bethe (1906–2005) at Cornell University when he and Bethe realized that the degenerate electrons in the interior of a white dwarf resemble the degenerate electrons in ordinary metals: they are excellent conductors of heat. For his Ph.D. thesis in 1939, Marshak calculated the thermal conductivity of the degenerate matter inside a white dwarf. Applying his results to Sirius B, Marshak found its highly degenerate interior to be very nearly isothermal, at a temperature of about 10 million degrees Kelvin (10^7 K) (Marshak 1940). At these temperatures—much lower than the billions of degrees obtained from radiative-transfer calculations—the electrons remain degenerate, so the models are self-consistent.

By the time World War II ended in 1945, another white-dwarf puzzle had come to the fore: what is the energy source that powers these dense stars? As we have seen, ordinary Main Sequence stars are powered by thermonuclear burning of hydrogen into helium. This cannot be the energy source for a white dwarf, however. The interior temperatures of these dense stars are about the same as the central temperatures of Main Sequence stars, but the densities are many orders of magnitude larger. If hydrogen-burning were the white-dwarf energy source, then their much higher densities would greatly increase the thermonuclear energy-generation rate, making the luminosity of a white dwarf much larger than the Sun's. However, the observed white-dwarf luminosities instead are many orders of magnitude smaller than the Sun's.

The answer to this puzzle was provided by British astrophysicist Leon Mestel (1927–2017). In 1952, he developed a simple cooling model for the evolution of a white dwarf (Mestel 1952). He assumed that essentially all of the heat inside the star is contained in the degenerate, isothermal core, which contains most of the stellar mass. The heat slowly leaks away through the very thin, non-degenerate surface layers of the white dwarf, which act like an insulating blanket around the hot core. Assuming that there are no other energy sources in a white dwarf, Mestel found that it simply cools slowly toward invisibility, the rate of cooling slowing as the star becomes fainter. While the time to cool to a luminosity of about one percent of the Sun's luminosity—fairly bright for a white dwarf—is about 100 million years, the cooling time to a luminosity a hundred times fainter may be several billion years or more.

The rate of cooling of a white dwarf depends upon the rate of heat transfer through its non-degenerate envelope. In turn, that depends upon whether energy is transported through the envelope by the flow of radiation, as Mestel had assumed for simplicity, or by convection. In his insightful small book on white dwarfs (Schatzman 1958), the French astrophysicist Evry Schatzman (1920–2010; see Figure 15.4) first pointed out that sufficiently cool white dwarfs must necessarily develop convection zones that extend inward for some distance below the stellar surface. A decade later, the German-born astrophysicist Karl-Heinz Böhm (1923–2014) included convection in constructing a detailed model for the outer layers of the white dwarf van Maanen 2. He found (Böhm 1968) that the efficiency of convective heat transfer significantly changes the relation between the luminosity of a white dwarf and the temperature of its degenerate, nearly isothermal core. This directly affects the cooling rates of these stars. The relatively thin subsurface convection zone also affects the observable stellar spectra and other properties of the white dwarf, as we shall see below.

15.1 The Spectra of White-dwarf Stars

The earliest photographic spectra obtained for white dwarfs during the opening decades of the 20th century showed similarities to the spectra of Main Sequence stars but also some striking differences. For example, Main Sequence stars classified as spectral type A exhibit strong absorption lines due to neutral hydrogen (HI lines[2]);

---

[2] Astronomers use the Roman numeral I to designate a neutral atom, with higher Roman numerals designating successive ionization stages.

Figure 15.4. Evry Schatzman in 1973. Credit: AIP Emilio Segrè Visual Archives, John Irwin Slide Collection; copyright American Institute of Physics.

somewhat hotter stars that display lines due to neutral helium (HeI) are assigned to spectral class B; and still-hotter stars that show lines of ionized helium (HeII) are classed as type O. The spectra of various white dwarfs exhibit similar patterns of spectral lines, and in analogy to the Main Sequence stars they are classified, respectively, as spectral types DA, DB, and DO, using the prefix "D" introduced by Luyten in 1945 to represent a degenerate star (Luyten 1945).

One obvious difference between white dwarfs and Main Sequence stars is the widths of the spectral lines. For Main Sequence stars, the widths of the HI lines at half the line depth are typically a few Ångstroms, but for white dwarfs the widths are often a few hundred Ångstroms. This difference is due to the much higher surface gravities of the white dwarfs—typically some 10,000 times greater than the surface gravity of the Sun—which produce much higher photospheric pressures that broaden the spectral lines.

A second peculiarity involves the abundances of the chemical elements in the stellar atmospheres (Wesemael et al. 1993). The spectrum of a DA white dwarf typically contains HI absorption lines and nothing else, while a DB white dwarf has HeI lines but no lines due to hydrogen or any heavier elements. In contrast, the atmospheres of Main Sequence stars produce absorption lines due to hydrogen *and* helium *and* the CNO elements *and* metallic elements like Si and Fe. The Fraunhofer spectrum of the Sun, for example, contains enormous numbers of absorption lines produced by all of these elements and more.

The plethora of white-dwarf spectral types, which are so different from Main Sequence spectral types, appears to be the result of several factors. In his book cited above, Schatzman pointed out that the high gravity of a white dwarf causes the heavier elements to sink rapidly out of sight below its photosphere. Thus, an atmosphere with an initial composition like that of the Sun would rapidly become a pure hydrogen (DA) atmosphere. And if the atmosphere should happen to be initially devoid of hydrogen, it would rapidly become a pure helium (DB or DO) atmosphere. In addition, some white dwarfs have spectra with no Main Sequence analogs. The DQ white dwarfs contain lines due to molecular and neutral carbon (C_2 and CI) but nothing else, while the DZ white dwarfs contain very weak lines due to metallic elements like Ca, Mg, or Fe but nothing else. And some white dwarfs—assigned to spectral class DC—have purely continuous spectra, with no lines deeper than about 5% of the continuum.

But if their high surface gravities cause the H in a white dwarf to float to the surface, why don't all white dwarfs have pure H spectra? In 1983, Icko Iben and his collaborators showed that the division between stars with surface layers that contain H and those that do not occurs naturally as a star leaves the planetary-nebula phase of stellar evolution (Iben et al. 1983). If such a star experiences a final thermal pulse, they found that the He-burning convective shell engulfs the residual H layer, and the H is completely burned. The star then swells once more to red-giant dimensions until He-burning is finally extinguished, ultimately returning to and proceeding down the white-dwarf cooling track. Iben subsequently showed that about 25% of the central stars of planetary nebulae go through this phase to become H-deficient white dwarfs —collectively termed "non-DA" white dwarfs—a result which he pointed out "is not in disagreement with the observations." (Iben 1984)

Additional factors appear to account for the trace amounts of heavy elements that appear in some white-dwarf spectra. At very high temperatures like those characteristic of the central stars of planetary nebulae, the intense radiation pressure produced by the extremely hot stellar core is able to levitate some heavy elements above the photosphere. At the other extreme, very cool white dwarfs may accrete heavy elements as they pass through clouds of interstellar gas and dust. Accretion onto a He-atmosphere white dwarf may be the explanation for the DZ stars. In between these two extremes of temperature, the subsurface convection zone extends deeper and deeper into the star as the white dwarf cools. At the same time, however, the mass contained within the degenerate core continues to increase as the star cools, pushing the degeneracy boundary (i.e., the boundary between the degenerate core and the non-degenerate envelope) outward toward the stellar surface. Just before the He convection zone and the degeneracy boundary meet, however, the convection zone may be able to extend deep enough into the star to dredge up some of the C from the outer reaches of the stellar core. This is thought to be the origin of the DQ white dwarfs.

15.2 Determining the Physical Properties of White Dwarfs

The spectrum of any star is determined by its surface gravity, effective temperature, and chemical composition. These physical properties can thus be determined for a given star by fitting synthetic spectra from sufficiently accurate theoretical model atmospheres to a sufficiently well-determined observational spectrum. Accurate values for the physical parameters obviously require both good theoretical models and good observational data. Both improved markedly over the last half of the 20th century.

In 1949, the enormous collecting area of the primary mirror of the 200-inch telescope made it possible to study much fainter astronomical objects than had previously been accessible. Caltech's Jesse L. Greenstein (1909–2002) immediately took advantage of this to obtain spectra of many hundreds of the white-dwarf candidates found by Luyten and others. He ultimately obtained spectra of more than 500 white dwarfs, many in collaboration with his colleague Olin J Eggen (1919–1998), which they published in a series of articles from the late 1950s (Greenstein 1958) into the mid-1960s. During a sabbatical leave from Universität Kiel in Germany to Caltech in 1962, Volker Weidemann (1924–2012; see Figure 15.5) capitalized on the availability of Greenstein's white-dwarf spectra to make the first serious attempt to use model atmospheres to determine the physical properties of the H-rich (DA) white dwarfs (Weidemann 1963). He obtained surface gravities for 22 DA stars from the profiles of their neutral hydrogen Hγ absorption lines, finding the average radius to be about 0.01 R_{Sun}—roughly the radius of the Earth—with the average white-dwarf mass being about 0.6 M_{Sun}. This work preceded the advent of high-speed digital computers, which first became widely available during the 1960s and which are essential for the substantial computational effort necessary to construct detailed model atmospheres and use them to analyze stellar spectra accurately.

Over the subsequent decades, spectroscopy evolved from photographic glass plates to vacuum-tube photomultipliers to solid-state devices, and the quantity and quality of white-dwarf spectra improved in consequence. At the same time, advances in computer technology enabled the creation of larger and faster programs capable of calculating increasingly detailed model atmospheres. By 1970, Harry L. Shipman, then a Caltech graduate student working with J. Beverly Oke (1928–2004), was able to use new spectrophotometric data obtained at the 200-inch telescope to show that the effective temperature of the DA white dwarf 40 Eridani B is $T_{eff} = 17,000$ K and

Figure 15.5. Volker Weidemann in 1980. Source: treywenger.com/rood_photos/erice80/. Photograph by the late Robert T. Rood.

that its radius is (0.0127 ± 0.0006) R_{Sun} (Oke & Shipman 1971). By the end of the 1970s, German astronomers Detlev Koester, H. Schulz, and Volker Weidemann used Koester's DA model atmospheres to carry out a pioneering analysis of high-precision narrow-band photometry that had been obtained by John Graham (1939–2018) of the Carnegie Institution of Washington (Graham 1972; Koester et al. 1979). For 86 DA white dwarfs, they obtained accurate values for T_{eff} and log g, the logarithm of the surface gravity, generally quoted in units of cm s^{-2}. Unexpectedly, they found that all of the stars clustered closely around log g = 7.88 ± 0.33, corresponding to a mean mass of about 0.58 M_{Sun}.

In the early 1990s, Pierre Bergeron and colleagues from the University of Arizona developed a χ^2-minimization technique to fit simultaneously all the observed HI Balmer lines of a DA white dwarf to the synthetic spectra obtained from Bergeron's own new DA model atmospheres. This approach yielded exceptional accuracy in determinations of the physical parameters of these stars. Bergeron and his colleagues employed this method to analyze 129 DA white dwarfs, obtaining a mean mass of 0.56 M_{Sun} (Bergeron et al. 1992), nearly identical to that found earlier by Koester and his colleagues. Beginning in 2000, the Sloan Digital Sky Survey (SDSS) began conducting a multispectral imaging and spectroscopic survey at the Apache Point Observatory in New Mexico (York et al. 2000). By 2006, both theory and observations had improved to the point where fits of the models to the spectroscopic data were nearly perfect (see Figure 15.6; Kepler et al. 2006). In consequence, astronomers were able to determine the white-dwarf mass distribution with some precision, finding the mean mass to be (0.65 ± 0.16) M_{Sun} (Giammichele et al. 2012).

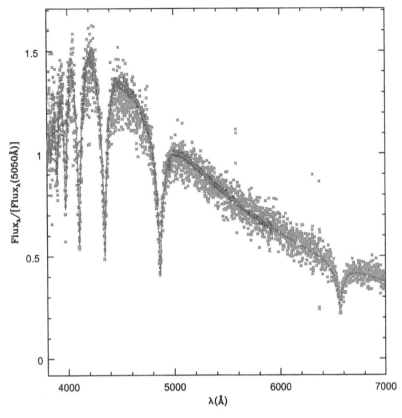

Figure 15.6. Fitting theoretical models to the observed Balmer-line profiles of the DA white dwarf J0303-0808 from Kepler et al. (2006), MNRAS, **372**, 1799; reproduced with permission from Oxford Univ. Press on behalf of the Royal Astronomical Society. The Gemini Observatory data are shown by the solid blue line, data from the Sloan Digital Sky Survey (SDSS) are shown as red crosses, and the theoretical model fits are shown by the dashed line, which is almost indistinguishable from the Gemini data. The resulting effective temperature is 11,960 K, $\log g = 8.305$, and the mass is 0.791 M_{Sun}.

The entire SDSS catalog contains almost 29,000 white dwarfs, the largest catalog of white-dwarf spectra to date.

Determining the luminosities of white dwarfs also requires knowledge of their distances, or equivalently, their parallaxes. By 1997, accurate trigonometric parallaxes had been obtained for 20 white dwarfs from the Hipparcos satellite (Vauclair et al. 1997), and almost 300 had been provided by the dedicated parallax program at the U.S. Naval Observatory (Dahn 1999). More recently, the Gaia satellite has provided parallax information, colors, and absolute magnitudes for more than 25,000 white dwarfs, enabling unprecedented comparisons of the observations to theory (Gaia Collaboration 2018). In 2018, Mukremin Kilic and his collaborators combined Gaia astrometry and photometry to show that the mass distribution of nearly 14,000 white dwarfs within 100 pc from the Sun is bimodal, with peaks in the mass distribution at about 0.6 M_{Sun} and 0.8 M_{Sun} (Kilic et al. 2018).

15.3 Pulsating White Dwarfs and Asteroseismology

Another chapter in the study of white dwarfs opened in 1968, when Louisiana State University astronomer Arlo Landolt serendipitously discovered oscillations with a period of about 12.5 minutes in the white dwarf star HL Tau 76 (Landolt 1968). His observations were soon explained by Brian Warner and Edward L. ("Rob") Robinson and by Landolt's LSU colleague Ganesh Chanmugam (1939–2004) as probable *g*-mode (buoyancy-based) oscillations of the white dwarfs (Warner & Robinson 1972; Chanmugam 1972). This set off extended efforts by observers to define the characteristics of oscillating white dwarfs and by theorists to understand and explain the oscillations.

High-speed photometric observations began to reveal more pulsating white dwarfs during the later 1960s and early 1970s. All were of spectral type DA—i.e., with pure H atmospheres—all displayed multiple oscillation periods, and all had effective temperatures in a narrow range from about 11,000 K to 12,500 K (Robinson 1979). In 1981, Donald E. Winget and his colleagues explained the oscillations of the pulsating ZZ Ceti white dwarfs[3] as *g*-modes trapped in the compositionally stratified surface layers of these stars (Winget et al. 1981). The stratification—which occurs to some extent in all white dwarfs—is produced by the high surface gravities of these dense stars, as pointed out decades earlier by Evry Schatzman. The oscillations occur in hydrogen-atmosphere white dwarfs only in a narrow range of effective temperatures between about 10,000 and 12,000 K because they are driven by changes that occur during an oscillation in the structure of the region of the star in which the ionization of H is incomplete. Winget also predicted that similar oscillations should occur in white dwarfs with helium atmospheres (classed as DB stars) but at higher effective temperatures because they are driven by the higher-temperature region in which He is incompletely ionized. This prediction was subsequently verified by Winget, by then at the University of Texas, Gilles Fontaine (1948–2019) from the Université de Montréal, and their collaborators at the McDonald Observatory in west Texas (Winget et al. 1982) when they discovered pulsations in the DB white dwarf GD 358.

Making sense of the multiplicity of periodicities in the white-dwarf variables required new developments in both observations and theory. Observationally, it was necessary to resolve the oscillation spectrum completely. This was challenging, because the amount of data obtainable from any single observatory inevitably contains half-day gaps during local daytime. A way to overcome this problem was pioneered by Rob Robinson, Ed Nather (1926–2014), and John McGraw in 1976, when they combined high-speed photometric observations from South Africa and Texas to resolve the oscillation spectrum of ZZ Ceti itself. Nather subsequently extended this approach to a concept he called the "Whole Earth Telescope" (Nather et al. 1990), in which cooperating observatories around the world participated in coordinated, extended observations of pre-selected target stars, producing

---

[3] ZZ Ceti stars—also called DAV stars because they are variable DA white dwarfs—are oscillating white dwarfs with pure hydrogen spectra.

observational records that may extend continuously for weeks at a time. Such long records have enabled astronomers to resolve completely as many as hundreds of distinct oscillation modes in some pulsating white dwarfs.

Analysis of the plethora of different modes has enabled astrophysicists to use theoretical model calculations to extract quantitative information about the physical properties of these stars from the observational data, an approach that has come to be called "asteroseismology." Asteroseismological determinations of the physical properties of white dwarfs are generally consistent with astronomical expectations, although there are a few exceptions. Masses tend to be about $0.6\ M_{Sun}$, the observational mean mass of the white dwarfs, with surface He-layer masses of about $10^{-2}\ M_{Sun}$ and overlying H-layer masses (for the DAVs) of about $10^{-4}\ M_{Sun}$. In 2018, Noemi Giammichele and her colleagues were able to use two years of asteroseismological observations from the Kepler spacecraft to probe the internal distribution of the chemical composition in the white dwarf KIC 08626021 (Giammichele et al. 2018). They resolved the observed oscillations of this star into eight g-modes, each with a period of some minutes, determined to microsecond precision. Because the relatively small number of g-modes in KIC 08626021 probe different depths within the star, Giammichele and her colleagues were able to place interesting constraints on the composition down through the surface layers and into the upper part of the C/O core (see Figure 15.7). These results are generally consistent with expectations based on stellar-evolution calculations, although the ratio of oxygen to carbon in this star is much greater than anticipated.

15.4 Magnetic White Dwarfs

Yet another part of the white-dwarf story began in 1970, when James C. Kemp and his colleagues discovered circular polarization in the spectra of some white dwarfs (Kemp et al. 1970). Detailed observations carried out by J. R. P. (Roger) Angel and John D. Landstreet, both then at Columbia University in New York City, revealed a number of white dwarfs with either circular or linear polarization (Angel & Landstreet 1970), which they found to be caused by extremely strong magnetic fields—with field strengths of hundreds of megagauss, more than 100 million times larger than the fields at the Sun's magnetic poles. The strong fields distort the spectrum of hydrogen—which proved to be present in most of the magnetic white dwarfs—making it completely unrecognizable. This highly distorted hydrogen spectrum in fact proved to be the explanation for the odd features Greenstein had discovered in the spectrum of one mysterious white dwarf, which he classified as "λ4135" for the wavelength of the strongest spectral feature.

About 10% of all white dwarfs have been found to contain magnetic fields with strengths exceeding a million gauss (one megagauss or 1 MG). Magnetic white dwarfs span the entire temperature range occupied by their non-magnetic siblings. The axis of the magnetic field is generally offset somewhat from the rotation axis, and the rotation periods range from 12 minutes to 18 days. Observational evidence suggests that most magnetic white dwarfs originated from the magnetic Ap and Bp stars on the Main Sequence. The white-dwarf fields have generally been thought to be "fossil" remnants of the original Main Sequence fields of these stars, with the

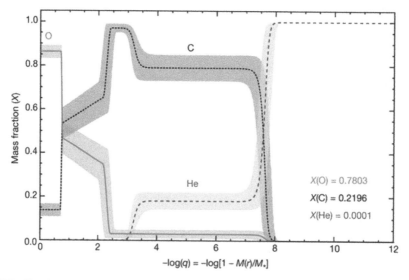

Figure 15.7. Chemical stratification in a white dwarf. From Giammichele et al. (2018). The vertical axis shows the mass fraction of the given element (all the mass fractions add up to unity), and the horizontal axis shows the logarithm of the fraction of the stellar mass measured inward from the stellar surface. In this plot, the left-hand edge corresponds to the center of the star, and the right-hand edge is within 10^{-12} M_{Sun} from the actual surface of the star. Thus, the outer layers are essentially pure He down to a depth of 10^{-8} M_{Sun} from the surface. Below this point (i.e., further to the left) the composition switches rather quickly to mainly C, with a mass fraction of about 20% of He and less than 5% of O. Between a depth of about 10^{-3} M_{Sun} down to less than 10^{-1} M_{Sun} (i.e., with a bit less than 90% of the mass of the star remaining down to the center), there is a complicated transition in the composition, and the remaining mass of the stellar core is about 85% O and 15% C by mass.

magnetic fields "frozen" into the stellar core that eventually became the white dwarf due to the high electrical conductivity of the core. Alternatively, the multi-megagauss magnetic fields observed in these dense stars may perhaps be generated by astrophysical dynamos like those in the Sun, planets, and other stars but much more powerful. Which of these ideas is correct remains to be determined.

15.5 The White Dwarf Luminosity Function, Core Crystallization, and Cosmochronology

The simple cooling theory that Leon Mestel published in 1952 (Mestel 1952) predicts the time required for a white dwarf to cool down to a given core temperature and luminosity. Realizing that this implies that the number of white dwarfs in a given range of luminosities increases as one proceeds to fainter and fainter white dwarfs (called the "white dwarf luminosity function" or WDLF), Volker Weidemann set out in 1963 to test this relationship observationally.[4] Because of the scarcity of data

---

[4] The luminosity function is a logarithmic plot of the numbers of white dwarfs in a given luminosity interval as a function of the luminosity. See Weidemann, V. 1967, ZA, **67**, 286, "Leuchtkraftfunktion und räumliche Dichte der Weißen Zwerge."

then available, about all that he could say was that the observations were more-or-less consistent with the theory; that is, there are more faint white dwarfs than bright white dwarfs per unit absolute magnitude.

As we have seen, the total number of known white dwarfs with reliable physical properties increased dramatically over the next half century (Fleming et al. 1986), particularly after the launch of the Gaia mission, and in 2019, Pier-Emmanuel Tremblay and his colleagues were able to produce a much more accurate plot of the WDLF, Φ, using Gaia data (see Figure 15.8; Tremblay et al. 2019). The red data points with error bars connected by dotted line segments in this figure are the actual Gaia data for white dwarfs with masses in the range 0.9–1.1 M_{Sun}, while the curves represent various theoretical models discussed below. In the range of luminosities $-1.7 \geqslant \log L/L_{Sun} \geqslant -2.8$, the data points correspond to a power law of the form $\log \Phi \approx -0.6 \times \log L/L_{Sun}$ plus a constant. For comparison, Mestel's theory gives $\log \Phi \approx -0.71 \times \log L/L_{Sun}$ plus a constant. The difference in the slope (-0.6 vs. -0.71) is pretty small, especially given the substantial advances in understanding the physical properties of white dwarfs over the past half century. This effectively confirms Mestel's cooling theory for the brighter white dwarfs.

However, the Gaia observations clearly depart from this straight-line fit at both higher and lower luminosities. What is going on there?

At the higher luminosities, the cooling times of the white dwarfs are decreased significantly by neutrino emission from the cores of these dense stars, as we saw in the previous chapter. The shorter cooling times mean that white dwarfs spend less time passing through a given range of luminosities than would be expected if the cooling were due solely to radiation from the stellar surface, as Mestel's theory

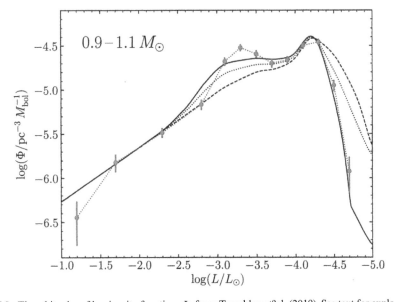

Figure 15.8. The white-dwarf luminosity function, Φ, from Tremblay et?al. (2019). See text for explanations of the curves.

assumes. However, the neutrino emission falls as the core of the white dwarf cools, and when the neutrino luminosity emitted by the core drops below the luminosity radiated from the surface, the cooling rate reverts to that given by Mestel's theory.

There is a bump in the WDLF in the range $-2.5 \geqslant \log L/L_{Sun} \geqslant -3.5$, which corresponds to another physical effect that was predicted in the 1960s. In 1961, Edwin E. Salpeter decided to investigate the effects of electrostatic interactions among the ions and electrons in a white dwarf (Salpeter 1961), which had been neglected in Chandrasekhar's pioneering calculations of the mechanical structure of idealized, zero-temperature white dwarf models. Salpeter found that these interactions decrease the pressure in a white dwarf by about 10%, which modifies the mass–radius relation for these stars (see Figure 15.3). Because of the electrostatic repulsion between the positively charged ions, he also concluded that at sufficiently low temperatures the lowest-energy state is one in which the ions form a crystalline lattice. A few years later, scientists at the Lawrence Livermore National Laboratory found that such a "one-component plasma" (OCP) undergoes a spontaneous transition from a fluid phase to a crystalline solid phase when the temperature drops below the freezing point of the OCP (Brush et al. 1966). At densities comparable to the central densities of white dwarfs, the crystallization temperature of the OCP turns out to be comparable to the central temperatures of these stars, with more massive white dwarfs freezing at higher temperatures. Since the fluid–solid transition is accompanied by a release of latent heat, crystallization temporarily slows the cooling rate, producing a pile-up of white dwarfs at a central temperature—and therefore luminosity—corresponding to this event (Van Horn 1968). Subsequent investigations of multicomponent plasmas—such as the C/O mixtures expected in white-dwarf cores—showed that phase separation occurs at the onset of crystallization, with the heavier, oxygen-rich phase "snowing out" into the deep interior while the lighter, C-rich fluid floats toward the surface, releasing additional gravitational energy (Mochkovitch 1983). At that time, however, there was no way to test these ideas about core crystallization in cooling white dwarfs.

The various theoretical curves in Fig. 15.9 correspond to various assumptions about core crystallization. The dashed curve neglects both the latent heat released during crystallization and the gravitational energy release that accompanies phase separation, with the heavier phase settling toward the center of the white dwarf. This curve clearly does not fit the Gaia data at all well. The dotted curve that does not pass through the data points includes the release of latent heat during core crystallization but still neglects the energy released during phase separation. Finally, the solid line includes both of these energy sources, and it fits the Gaia data points better than the other theoretical curves. While this confirms the existence of core crystallization in sufficiently cool white dwarfs, the remaining differences show that we still have things to learn about this process.

Finally, for luminosities fainter than $\log L/L_{Sun} \leqslant -4.2$, there is a sharp drop-off in the WDLF. This turns out not to be caused by an internal physical property of the white dwarfs but instead is a consequence of the finite age of the disk of the Milky Way Galaxy. The first indication of this drop-off came in the late 1970s from a dedicated search for extremely faint white dwarfs by Jim Liebert and his Arizona colleagues (Liebert et al. 1979). They concluded that there was probably a real deficiency of these faint stars, and they pointed out that this is consistent with the idea that star formation in the galactic disk began less than about 10 billion years (10 Gyr) ago. Continuing advances in both observation and theory have reinforced this idea, and less than a decade later Don Winget and his colleagues used the best white-dwarf models then available to conclude that the age of the galactic disk is 9.3 ± 2.0 Gyr (Winget et al. 1987). In 1994 Laurent Segretain, Margarita Hernanz, and their European colleagues pointed out that not only C/O phase separation but also the gravitational segregation of trace elements such as ^{22}Ne and ^{56}Fe, which remain from earlier stages of stellar evolution, can release significant additional energy (Segretain et al. 1994). This can extend the age estimate for the galactic disk obtained from white-dwarf cosmochronology—as this technique has come to be called—by 1.5 to 2 billion years. Such additional energy sources may also affect the magnitude and shape of the core-crystallization bump in the WDLF. All the theoretical models plotted in Fig. 15.9 assume a disk age of 10 Gyr.

References

Angel, J. R. P., & Landstreet, J. D. 1970, ApJ, 160, L147

Bergeron, P., Saffer, R. A., & Liebert, J. 1992, ApJ, 394, 228

Böhm, K.-H. 1968, Ap&SS, 2, 375

Brush, S. G., Sahlin, H. L., & Teller, E. 1966, JCP, 45, 2102

Chandrasekhar, S. 1935, MNRAS, 95, 207

Chanmugam, G. 1972, NPhS, 236, 83

Dahn, C. C. 1999, 11th European Workshop on White Dwarfs, ASP Conf. Ser. Vol. 169 (San Francisco, CA: ASP) 24

Eddington, A. S. 1926, The Internal Constitution of the Stars (Cambridge: Cambridge Univ. Press) 171

Fleming, T. A., Liebert, J., & Green, R. F. 1986, ApJ, 308 176; see also
    Liebert, J., Dahn, C. C., & Monet, D. G. 1988, ApJ, 332, 891

Gaia Collaboration 2018, A&A, 616, A10

Giammichele, N., et al. 2018, Natur, 554, 73

Giammichele, N., Bergeron, P., & Dufour, P. 2012, ApJS, 199, 29

Graham, J. A. 1972, AJ, 77, 144

Greenstein, J. L. 1958, HDP, 11, 161

Hamada, T., & Salpeter, E. E. 1961, ApJ, 134, 683

Iben, I. Jr. 1984, ApJ, 277, 333

Iben, I. Jr., Kaler, J. B., Truran, J. W., & Renzini, A. 1983, ApJ, 264, 605

Kemp, J. C., Swedlund, J. B., Landstreet, J. D., & Angel, J. R. P. 1970, ApJ, 161, L77

Kepler, S. O., et al. 2006, MNRAS, 372, 1799

Kilic, M., et al. 2018, MNRAS, 479, L113

Koester, D., Schulz, H., & Weidemann, V. 1979, A&A, 76, 262

Kuiper, G. P. 1941, PASP, 53, 248

Landolt, A. U. 1968, ApJ, 153, 151

Liebert, J., et al. 1979, ApJ, 233, 226

Luyten, W. J. 1945, ApJ, 101, 131

Marshak, R. E. 1940, ApJ, 92, 321

Mestel, L. 1952, MNRAS, 112 583; see also
  Mestel, L., & Ruderman, M. A. 1967, MNRAS, 136, 27

Mochkovitch, R. 1983, A&A, 122 212; see also
  Salaris, M., et al. 1997, ApJ, 486, 413

Nather, R. E., et al. 1990, ApJ, 361, 309

Oke, J. B., & Shipman, H. L. 1971, IAUS 42 White Dwarfs, ed. W. J. Luyten (Dordrecht: Springer) 67

Robinson, E. L. 1979, Proc. IAU Colloq. 53: White Dwarfs and Variable Degenerate Stars, ed. H. M. Van Horn, & V. Weidemann (Rochester, NY: Univ. Rochester) 343

Salpeter, E. E. 1961, ApJ, 134, 669

Schatzman, E. 1958, White Dwarfs (Amsterdam: North Holland)

Segretain, L., et al. 1994, ApJ, 434 641; see also
  Hernanz, M., et al. 1994, ApJ, 434, 652

Tremblay, P.-E., et al. 2019, Natur, 565, 202

Van Horn, H. M. 1968, ApJ, 151, 227

Vauclair, G., et al. 1997, Hipparcos Venice'97, ed. B. Battrick (Noordwijk: ESA) 371

Warner, B., & Robinson, E. L. 1972, NPhS, 234, 2

Weidemann, V. 1963, ZA, 57, 87

Wesemael, F., et al. 1993, PASP, 105, 761

Winget, D. E., Robinson, E. L., Nather, R. D., & Fontaine, G. 1982, ApJ, 262, L11

Winget, D. E., Van Horn, H. M., & Hansen, C. J. 1981, ApJ, 245, L33

Winget, D. E., et al. 1987, ApJ, 315, L77

York, D. G., et al. 2000, AJ, 120, 1579

Chapter 16

The Evolution of High-Mass Stars and Supernovae

The evolution of a high-mass star begins on the Main Sequence with the thermonuclear burning of H into He, just as it does with any other star. High-mass stars differ in very significant ways, however. For one thing, the temperature distribution throughout a Main Sequence star increases as we proceed to larger stellar masses. Consequently, radiation pressure becomes increasingly important, as Eddington showed more than a century ago (*cf.*, Chapter 3). The rate of H-burning necessary to supply the star's luminosity also increases with the stellar mass, and the higher central temperatures mean that CNO reactions increasingly dominate over p–p chain reactions. The nuclear-burning rates in these stars are so large that their cores become convective, and H is therefore converted into He uniformly over the entire convective core. Because the stellar luminosity increases dramatically with the stellar mass, as shown for example by Eddington's mass–luminosity relation (see Chapter 3), the timescales for the post-Main-Sequence evolution of massive stars are much shorter than they are for lower-mass stars. Because of its high luminosity, the evolution of a high-mass star is dominated by stellar winds and the presence of numerous convection zones. Unfortunately, because there is no first-principles theory for either convection or mass loss, they can only be treated phenomenologically.

The minimum mass for stars classified as "high mass" is about 8 M_{Sun}, which is approximately the maximum mass that can produce a white dwarf. A number of publications over the past two decades have described our current understanding of the evolution of such stars in considerable detail (Heger *et al.* 2003). Such stars ignite He-burning at central temperatures of about 100 million degrees Kelvin (10^8 K). At this temperature, the star undergoes thermonuclear burning of He into C and O by means of the triple alpha (3α) reaction, and a subsequent alpha-capture by the newly formed ^{12}C nucleus forms ^{16}O. Meanwhile, H continues to burn into He in a shell surrounding the He layer, gradually depleting the outer H envelope and adding to the mass of He, while He continues to burn into C and O in the central core.

Beyond the end of central He-burning, the core of a high-mass star evolves much more rapidly than does the envelope due to the increasing importance of neutrino energy losses from these high-temperature phases of evolution, so the appearance of the star in the H–R diagram does not reflect the evolution of its central regions (Ekström 2021). Stars with initial Main Sequence masses greater than 9 M_{Sun} pass through several additional nuclear-burning stages. Because a high-mass star is too large ever to be supported by the pressure of degenerate electrons, the evolution of such a star consists of a series of thermonuclear-burning episodes, with gravitational contraction of the "ashes" of each burning stage heating them up until the central temperature becomes high enough to ignite the next stage of nuclear burning. When central He-burning has been completed, for example, the core contracts again, heating to sufficiently high temperatures (of the order of 600–800 million degrees Kelvin) for C to undergo further nuclear-fusion reactions to form still heavier elements, mainly ^{16}O, ^{20}Ne, and ^{23}Na (Clayton 1968a). Ultimately, rapid electron captures lead to the collapse of the ONeMg core, resulting in a supernova explosion that leaves behind a neutron star remnant (Heger *et al.* 2003; Janka 2012).

After C burning, Ne burning occurs before O burning because ^{16}O is very stable (Ekström 2021). The first major Ne-burning reaction, which occurs at a temperature of about 1.6×10^9 K, is the photodisintegration of ^{20}Ne: $\gamma + {}^{20}$Ne $\rightarrow$ ^{16}O $+ \alpha$. The net effect of Ne burning is to convert two ^{20}Ne nuclei into ^{16}O $+ {}^{24}$Mg. At the end of Ne burning, the core consists mainly of ^{16}O, ^{24}Mg, and ^{28}Si. The core then contracts again to about 2×10^9 K, where O burning occurs (Ekström 2021). Most of the energy liberated during O burning is radiated away as neutrinos (Clayton 1968b). At the end of O burning, the core consists mainly of ^{28}Si and ^{32}S.

The Coulomb barrier between the nuclei is too large for the reaction ^{28}Si $+ {}^{28}$Si to proceed. Instead, the energy released during Si burning comes mainly from photodisintegrations at temperatures greater than 3×10^9 K. The Si-burning reactions build a distribution of the nuclides in statistical equilibrium all the way up to ^{56}Fe (Ekström 2021). This final stage of Si burning lasts about two weeks, and it produces a core of about 1.5 M_{Sun} of iron-peak elements (Woosley & Janka 2005). At this point, the star has an "onion-skin" compositional structure (see Figure 16.1), with a core of the iron-peak elements Fe, Ni, and Co overlain by a layer of silicon-rich matter, overlain by a layer of O, Ne, and Mg, overlain by a layer of C and O, then a layer of He, and finally a surface H-rich layer consisting of material that has not undergone thermonuclear burning. At the base of each compositional layer is a thermonuclear-burning shell that continues to process matter from the overlying layer into the composition directly below it. In the process, a star with an initially uniform composition dominated by hydrogen develops the layered structure described above, as pointed out in 1959 by Alistair G. W. Cameron (1925–2005) (Cameron 1959). At the same time, the luminosity of the star increases more-or-less steadily to higher and higher luminosities, while the star's center becomes progressively denser and hotter. Because ^{56}Fe is the most stable of all nuclides,

Figure 16.1. Schematic illustration of the "onion-skin" structure of the chemical composition inside a high-mass star near the end of nuclear burning in the stellar core. At the base of each compositional shell there is a nuclear-burning zone that processes the composition of the upper layer into the heavier elements in the next shell. From commons.wikimedia.org/wiki/File:Evolved_star_fusion_shells.svg.

additional energy cannot be released by fusing two ^{56}Fe nuclei together. The stage is now set for the final phase of the evolution of a high-mass star.

16.1 Collapse of the Iron Core

The formation of a core of iron-peak elements is the precursor to a catastrophic stellar explosion caused by the collapse of the core. To support the continuing loss of energy from the core—by thermal conduction to the outer layers, neutrino energy losses directly from the core, electron captures onto nuclei, which reduce the pressure support from degenerate electrons, and photodisintegrations (Woosley & Janka 2005)—the iron core must continue to contract, releasing gravitational potential energy and becoming denser and hotter as it approaches the limiting mass that can be supported by electron degeneracy. When the core mass exceeds this limit, it collapses, generally forming either a neutron star (see Chapter 18) or a black hole (Chapter 19), depending upon the mass of the presupernova star (Janka 2012). The collapse from an iron core about the size of Earth to a proto-neutron star with a radius of about 30 km occurs in a fraction of a second (Janka 2012). This is much faster than the relevant timescale for the rest of the star, which may be hours to as much as a day. Consequently, the imploding core becomes decoupled from the rest of the star, which continues to collapse inward onto the core over much longer timescales.

As the density of the core increases, it becomes energetically favorable for the degenerate electrons (which have been primarily responsible for supporting the overlying layers of the star against gravity) to be captured into the atomic nuclei. This essentially converts the protons in the nuclei into neutrons, emitting a neutrino in the process. The nuclei become increasingly neutron enriched—a process called "neutronization"—and above a density of about 4×10^{11} g cm^{-3} the nuclei have become so neutron-enriched that free neutrons begin to "leak" out of the nuclei. The additional free nucleons do not exert sufficient additional pressure to halt the collapse, however, and it continues all the way to nuclear densities, about 2×10^{14} g cm^{-3}. Above this density, the remaining nuclei "dissolve" into a "sea" of free neutrons, together with a "seasoning" of free protons and electrons. Within a millisecond, the strong repulsion due to the hard cores of the nucleons finally halts and reverses the collapse. A proto-neutron star with a mass of about 1.4 M_{Sun} has been formed, and the end of the collapse of these central regions sends a shock wave out into the still-collapsing outer part of the stellar core.

The amount of gravitational energy released in the formation of a neutron star is more than 10^{53} ergs, and because it is released in a minute fraction of a second this is like setting off an unimaginably stupendous bomb in the center of the star. The total amount of energy released is in fact orders of magnitude larger than the amount of energy necessary to disrupt the entire mass of the star remaining above the newly formed neutron star. For example, the gravitational binding energy of a 30 M_{Sun} Main Sequence star is about 5×10^{50} ergs, one thousand times less than this. A stellar explosion of such tremendous magnitude was first termed a "supernova" by astronomers Walter Baade (1893–1960) and Fritz Zwicky (1898–1974) in 1934 (Baade & Zwicky 1934). As they remarked, "at their maximum brightness they emit nearly as much light as the whole nebula [N.B., external galaxy] in which they originate." (For example, see Figure 16.2 from Soderberg et al. 2008). Baade and Zwicky estimated the total amount of radiation emitted during a supernova outburst to be of the order of 10^{48} ergs at visible wavelengths alone. This is as much energy as the Sun radiates at all wavelengths in eight million years.

Astrophysicists initially thought that the shock wave produced by the formation of the neutron star was responsible for causing the outer layers of the progenitor star to explode. However, the burst of neutrinos emitted during the early stages of neutronization carries away so much energy that the shock wave stalls. But as the outer parts of the core continue to fall inward, the matter just above the surface of the proto-neutron star becomes dense enough to trap some of the elusive neutrinos within this inward flow, and the energy deposited there by the neutrinos appears to be sufficient to trigger the explosion.

The Crab Nebula in the constellation Taurus (see Figure 16.3) is a remnant of such a supernova, which exploded in 1054 CE and which was observed widely

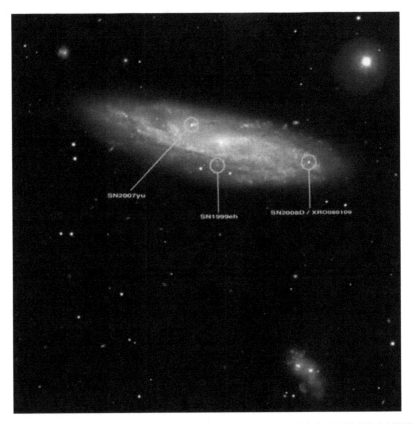

Figure 16.2. Supernovae in the galaxy NGC 2770: SN 2007yu, SN 1999eh, and SN 2008D/XRO 080109, from: https://noirlab.edu/public/images/gemini0804a. SN 2008D is a Type Ibc core-collapse supernova, which originated from a star with a mass greater than about 20 M_{Sun} that had lost its H-rich envelope. It produced a kinetic energy of about 10^{51} ergs and created a significant fraction of a solar mass of ^{56}Ni that powered the latter part of its light curve.

around the world (Mayall 1937). At its core, the Crab Nebula contains a rapidly rotating neutron star with an extremely strong magnetic field, perhaps a hundred billion times the strength of the Sun's magnetic field. Rotating about 30 times per second, this object emits regular pulses of optical, X-ray, and radio radiation each time the magnetic poles swing past the Earth. Surrounding this small, dense "pulsar," as it is called, is the tenuous, expanding gas cloud that is the Crab Nebula—the gaseous remnant of the disrupted progenitor star—which glows at optical and radio wavelengths from the re-emission of energy pumped into it by the pulsar.

Finally if the supernova progenitor is sufficiently massive, the rest of the stellar matter falling onto the proto-neutron star can push it beyond the limit of stability for neutron stars (analogous to the Chandrasekhar limit for white dwarfs that are supported by the pressure exerted by degenerate electrons, but in this case due to the pressure exerted by degenerate neutrons), causing the proto-neutron star to continue to collapse into a black hole.

Figure 16.3. The Crab Nebula. From en.wikipedia.org/wiki/File:Crab_Nebula.jpg; courtesy NASA and ESA.

16.2 Supernova 1987A

In February 1987, Ian Shelton was the resident astronomer at the Las Campanas Observatory in Chile. On the 24th of that month, he took a 3-hour photographic exposure of a section of the southern sky that included the Large Magellanic Cloud (LMC)—a small galaxy that is a satellite of our Milky Way Galaxy—planning to search for novae. When he developed the image and compared it with one from the night before, he found an extra star in the new image. It had been a faint star the night before, but that night the same object was a bright star (Livingstone 2013). He and other astronomers quickly realized that the new object was in fact a supernova—designated SN 1987A, the first to be seen in the vicinity of our own galaxy since Kepler's supernova in 1604—and they announced the discovery to the world in Circulars from the International Astronomical Union (Marsden 1987a).

Astronomers, astrophysicists, space scientists, and experimental physicists from around the world immediately trained every available observing device on the new supernova. The possible progenitor was an OB star named Sandulaek −69°202 (Marsden 1987b), which indicates that SN 1987A is in fact located within the LMC. Observers monitored the changes in the photometric colors and the spectrum of the supernova using ground-based telescopes and space observations (Marsden 1987c). By 1990, sufficient observational evidence had accumulated to confirm that radioactive decay of 0.075 M_{Sun} of ^{56}Ni into ^{56}Co was indeed the energy source for the initial decline in the light curve of SN 1987A (Suntzeff & Bouchet 1990), while the subsequent decay of ^{56}Co powered the decline over the first couple of years after the outburst. The decay of about 0.01 M_{Sun} of ^{57}Co provided the energy during the subsequent decline (see Figure 16.4).

The new supernova offered an unprecedented opportunity to check theoretical models for core-collapse supernovae. Since the 1960s, theorists had recognized that in every core-collapse supernova, the process of neutronization releases a veritable flood of neutrinos. Because they interact so weakly with ordinary matter, neutrinos escape rapidly from the deep interior of the collapsing core. Matter at densities like that inside the Sun are essentially transparent to neutrinos, and they escape from the solar core almost unimpeded, as we saw in Chapter 8. At densities as high as those in the collapsing core of a supernova, however, matter is no longer transparent to neutrinos, but even so they diffuse out of the dying star in a matter of seconds.

Physicists had already constructed a few neutrino detectors to investigate the physical properties of these ghostly particles. In 1981, a collaboration between the University of California at Irvine, the University of Michigan, and Brookhaven National Laboratory had constructed one—a huge tank filled with 5000 metric tons of purified water, called the "IMB detector"—600 meters (more than a third of a mile) underground in a Morton Salt Mine on the shore of Lake Erie. A few years

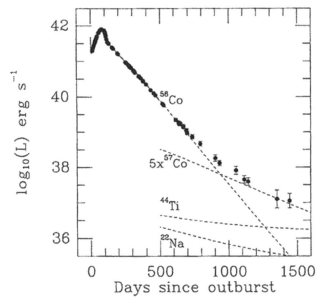

Figure 16.4. Light curve of SN 1987A. The vertical axis gives the logarithm of the luminosity, and the horizontal axis gives the number of days since the outburst. The various dashed lines show the exponential decays of the radioactive nuclides ^{56}Co, ^{57}Co, ^{44}Ti, and ^{22}Na. From Suntzeff *et al.* (1992).

later, physicists in Japan constructed a similar detector located 1000 meters underground in the Mozumi Mine near Kamioka in Japan. The original 3000-ton detector, called "Kamiokande," was completed in 1983. It was upgraded in 1985, and Kamiokande II took data from 1985 to 1990. Both the IMB and Kamiokande II detectors were in operation when SN 1987A exploded, and both detected neutrinos from the supernova, which they reported in companion papers in *Physical Review Letters* (Hirata *et al.* 1987); Kamiokande II detected 11 events during a 13-second burst on 1987 February 23, while the IMB detector observed eight events during 6 seconds preceding the optical detection of SN 1987A.

Although fewer than 20 neutrinos were detected, theorists immediately began to investigate the implications. One group concluded that the formation of a black hole could be ruled out and that the neutron star formed by the collapsing core of SN 1987A has a mass lying in the range 1.0–1.8 M_{Sun} (Sat & Suzuki 1987).

Of course, matter from the outer parts of the supernova progenitor, which was ejected during the explosion, has continued to expand in the decades since, and by 2019, it was observed by the Atacama Large Millimeter–Submillimeter Array (ALMA) in northern Chile as a ring of material almost $2''$ in diameter, with a warm central "blob," possibly heated by a "compact source" (Cigan *et al.* 2019). The most plausible source is in fact the neutron star formed during the explosion (termed NS 1987A), and models for cooling neutron stars fit the observations as well as observations of the next-youngest neutron star, in the supernova remnant Cassiopeia A (Page *et al.* 2020).

16.3 Other Types of Supernovae

Watchers of the skies have noted the appearance of the objects we now call supernovae—along with ordinary novae, comets, and other celestial changes—since time immemorial (https://en.wikipedia.org/wiki/History_of_supernova_observation), but the modern interest in supernovae can be dated to the careful observations of Danish astronomer Tycho Brahe (1546–1601). In 1572, he discovered a new star where none had previously existed near the constellation Cassiopeia (see Figure 16.5). This was a truly momentous discovery, because Brahe had spent years observing the heavens with great care and measuring the positions of celestial objects with the greatest precision of which the best astronomical devices of the day were capable, so he was as familiar with the heavens as a parent is with the face of his or her child. He used the Latin phrase *"stella nova"* (literally, "new star") to describe the abrupt appearance of this previously unknown object[1]. Continuing observations over the following months found that the brightness of Tycho's new star faded gradually, until it finally became invisible. It is known today as the supernova of 1572 or "Tycho's supernova." Three decades later, in 1604, Brahe's German contemporary Johannes Kepler (1571–1630) observed another supernova, which behaved in a similar manner.

---

[1] I am grateful to Dr. Joseph Pesce, of the Division of Astronomical Sciences at the National Science Foundation, for a number of interesting discussions about this historical point.

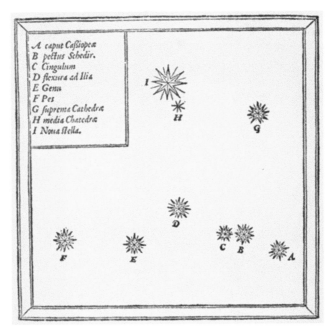

Figure 16.5. Tycho's supernova of 1572 in Cassiopeia, showing the location of the supernova amid the other stars in the constellation Cassiopeia. The supernova is the star labelled "*I Nova Stella.*" This is a photocopy of a drawing from Tycho Brahe's book *De nova stella* published in 1573. Source: commons.wikimedia.org/wiki/File:Tycho_Cas_SN1572.jpg.

Because supernovae occur so infrequently—in our own Milky Way Galaxy, they are estimated to occur about twice per century on average—it was not possible to study them in any detail until the advent of telescopes with large apertures (e.g., the 100-inch Hooker Telescope on Mt. Wilson) made it possible to study supernovae in distant galaxies (Branch & Wheeler 2017). In 1941, the German-American astronomer Rudolph Minkowski (1895–1976) classified supernovae spectroscopically into Types I and II, depending upon whether their spectra did not or did contain H lines. He found that nine supernovae of Type I (SNe I) formed a very homogeneous group: they exhibit *no* hydrogen lines (Minkowski 1941). The remaining five objects in his limited sample he classified as supernovae of Type II (SNe II). Their spectra were distinctly different from those of Type I, and they all displayed prominent features due to hydrogen.

This spectral classification of supernovae has since been further subdivided (Da Silva1993). Supernovae of Type I are classified into subtypes Ia, which exhibits a line due to ionized Si near peak light; Ib, which displays no Si but does exhibit neutral He; and Ic, which shows neither Si nor He (https://en.wikipedia.org/wiki/Supernova). Similarly, while all SNe II exhibit H lines, those of Type II-P have a plateau in their light curves; in Type II-L the visual magnitude decreases linearly with time; Types IIn display some narrow lines; and those of Type IIb change to resemble Type Ib.

The light curves of supernovae display a bewildering variety of different shapes (see Figure 16.6). All are characterized by a relatively rapid rise to maximum light

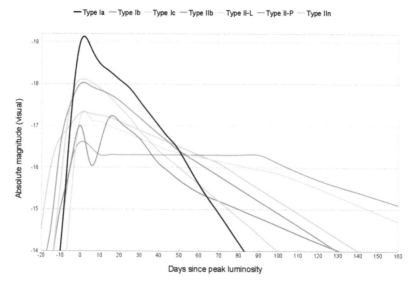

Figure 16.6. Comparative supernova light curves. The type of supernova is indicated by the color key at the top of the figure. From: commons.wikimedia.org/wiki/File:Comparative_supernova_type_light_curves.png.

(in a time of days to weeks), a duration near peak luminosity of weeks to about a month, and a protracted decline in brightness over a significant fraction of a year. As Figure 16.6 shows, the luminosity of a supernova at its brightest is a significant fraction of the total power output of an entire galaxy containing perhaps a hundred billion stars. As noted above, such an enormous release of energy is sufficient to disrupt a star completely, spewing the gases of which it had been composed back out into space. The explosion leaves behind an expanding gaseous nebulosity. Astronomers have in fact detected the remnant left behind by the explosion of Tycho's supernova, as well as many others (https://en.wikipedia.org/wiki/List_of_supernova_remnants).

In 1960, Fred Hoyle (1915–2001) and William A. Fowler (1911–1995) considered the physical causes of supernova explosions in some detail (Hoyle & Fowler 1960). Assuming thermonuclear fusion to be the energy source of a supernova explosion, they pointed out that in order to cause an explosion, the nuclear fuel must be capable of releasing its energy on a timescale of the order of 1–100 seconds. They argued that the fuel cannot be H or He, because the weak interactions involved in H-burning take too long, and the isotope ^{8}Be involved in the 3α reaction are never present in significant concentrations. However, they concluded that ^{12}C, ^{16}O, and other elements up to Si are all sufficiently explosive at temperature of about 1.5 to 2.5 billion Kelvin (1.5–2.5×10^9 K) and that the thermonuclear burning of only 0.1 M_{Sun} of these elements releases about 10^{50} ergs of energy. This is more than enough to account for the observed energy release in a supernova explosion.

Hoyle and Fowler further concluded that a catastrophic stellar explosion can take place in one of two different circumstances: (1) If the nuclear fuel is non-degenerate, a catastrophic implosion of the stellar core is necessary to trigger the explosion of the outer layers of a star. This happens in massive stars (say, a star of 30 M_{Sun}), and they identified the resulting explosion as a supernova of Type II. We have already discussed this type of supernova above. (2) On the other hand, if the nuclear fuel is degenerate, it is inherently unstable because the pressure balance of degenerate matter is only weakly affected by the heating that accompanies energy release. Consequently, a catastrophic explosion does not require a core collapse as a trigger. Such conditions can in principle occur in a star with a mass only slightly more than 1 M_{Sun}, and they identified the resulting explosion as a supernova of Type I.

16.4 How Does a Type I Supernova Explode?

The conclusions drawn by Hoyle and Fowler point to massive white dwarfs as the progenitors of Type I supernovae. We have seen how normal stellar evolution produces white dwarfs, and some of the observed white dwarfs have masses of 1 M_{Sun} or more. However, every white dwarf less massive than the Chandrasekhar limit of about 1.4 M_{Sun} is completely stable. So how can a massive white dwarf be triggered into instability? John Whelan (1945–1982) and Icko Iben Jr. considered this problem in 1973 and concluded that the answer may involve a close binary system containing a massive C/O white dwarf primary and a 0.8 M_{Sun} red giant (Whelan & Iben 1973). When the companion star fills its Roche lobe and transfers mass onto the primary, the white dwarf may approach the Chandrasekhar limit, experiencing sufficiently rapid contraction and heating to ignite a C-burning deflagration or detonation, resulting in a Type I supernova explosion. However, exactly this type of configuration produces cataclysmic variables, and nova eruptions—one type of cataclysmic variable—expel all the accreted mass and perhaps more (see Chapter 21). So how can the white dwarf accumulate sufficient mass to push it over the limit?

Two possibilities have been suggested. In one, dubbed the "single-degenerate" (SD) scenario, the white dwarf actually does grow in mass. This can occur if the accretion rate is sufficiently high. Calculations of accretion onto hot, luminous, and massive C/O white dwarfs show that not only is H burned steadily into He, but also that as the He layer grows in mass, it gradually fuses into C and O at its base. In contrast to nova calculations, which involve accretion onto much cooler and less luminous white dwarfs and where H is burned in a thermonuclear outburst and mass is ejected, during accretion onto such hot, luminous white dwarfs, the H burns as it is accreted and the white dwarf consequently grows in mass (Starrfield *et al.* 2004), ultimately pushing the mass of the white dwarf to the Chandrasekhar limit. Alternatively, once the H-rich envelope of the companion star has been lost, the residual He core of the companion may begin to transfer He directly onto the white dwarf (Warner 1974). As its mass approaches the Chandrasekhar limit, the white dwarf contracts, releasing gravitational energy that slowly heats the star. Carbon-burning may then begin, at first "simmering" in the core, until the temperature rises sufficiently to trigger a thermonuclear runaway, and the star literally explodes.

The other possibility, dubbed the "double-degenerate" (DD) scenario, assumes that two white dwarfs with masses below the Chandrasekhar limit are in a close binary orbit. If the orbit contracts—due, e.g., to gravitational radiation—the two white dwarfs eventually collide, and the energy released produces a supernova explosion (Webbink *et al.* 1984). The SD and DD scenarios each have their partisans, but as of the time of this writing, the SD scenario seems to be favored, with the question of whether or not the DD scenario can lead to a Type Ia supernova remaining an open question (Raskin *et al.* 2012).

Following the supernova explosion, the light curves of SNe Ia decline approximately exponentially with a half-life of about 55 days, although the half-life ranges from less than 42 days to more than 87 days (Mihalas 1963). Early supernova models had significant difficulty in producing the observed amount of optical emission, because most of the energy goes into the kinetic energy of the expanding gaseous remnants rather than into radiation. Powering the observed supernova light curve thus requires the release of energy during the expansion of the remnant. This originally led Geoffrey R. Burbidge (1925–2010) and his colleagues to suggest that the light curves might be powered by the spontaneous fission of ^{254}Cf (Burbidge *et al.* 1956), which had been discovered only about five years before and which has a half-life of 60.5 days (https://en.wikipedia.org/wiki/Californium). However, it is not clear how a sufficient amount of ^{254}Cf could be produced. In contrast, the high temperatures achieved during a SNe Ia explosion not only are high enough to burn C and O, producing a composition dominated by Si and S, but—as pointed out by James W. Truran (1940–2022) in the late 1960s and employed by Stirling A. Colgate (1925–2013) and Chester McKee (1942–2008) in 1969—also to burn Si, resulting in a composition initially dominated by ^{56}Ni (Colgate & McKee 1969). With a half-life of 6.077 days, the radioactive decay of about 0.25 M_{Sun} of ^{56}Ni to ^{56}Co is sufficient to supply the observed emission of radiation near the peak luminosity of the supernova; this implies that much of the mass of the original C/O white dwarf has been converted into ^{56}Ni. The subsequent decay of ^{56}Co into ^{56}Fe takes 77 days and is responsible for powering the tail of the supernova light curve (Leventhal *et al.* 1980).

Supernovae of Type Ia are of particular interest because the shapes of their light curves are very similar, and the total amount of energy released appears to be almost the same, corresponding to the energy that would have to be released to disrupt completely a white dwarf with a mass of about 1.4 M_{Sun}—essentially the Chandrasekhar limiting mass for a white dwarf. Because of this uniformity in the total energy release, SNe Ia were employed as "standard candles" to measure the distance to very remote galaxies (those with large cosmological redshifts) by Saul Perlmutter and his colleagues in their Nobel-Prize-winning discovery that the expansion of the universe is accelerating.

Some unresolved issues remain, however. As of this writing, it is still not clear whether the SD or the DD scenario is correct or whether SNe Ia involve some combination of the two. It is also not entirely clear how the initial stages of the supernova eruption proceed. Does carbon burn in a deflagration (essentially a subsonic thermonuclear "flame") or in an explosive detonation? Or does an initial deflagration turn into a detonation after some delay? Or is some other "non-standard" process responsible? (Khokhlov *et al.* 2000).

References

Baade, W., & Zwicky, F. 1934, PNAS, 20, 254

Branch, D., & Wheeler, J. C. 2017, Supernova Explosions (Berlin: Springer Nature); see also the excellent summary in https://en.wikipedia.org/wiki/Supernova

Burbidge, G. R., et al. 1956, PhRv, 103, 1145

Cameron, A. G. W. 1959, MSRSL 5th ser., 3, 163

Cigan, P., et al. 2019, ApJ, 886, 51

Clayton, D. D. 1968a, Principles of Stellar Evolution and Nucleosynthesis (New York: McGraw-Hill) 430

Clayton, D. D. 1968b, Principles of Stellar Evolution and Nucleosynthesis (New York: McGraw-Hill) 433

Colgate, S. A., & McKee, C. 1969, ApJ, 157, 623

Da Silva, L. A. L. 1993, Ap&SS, 202, 215

Ekström, S. 2021, FrASS, 8, 53

Heger, A., et al. 2003, ApJ, 591 288; see also
    Woosley, S. E., & Janka, H. T. 2005, NatPh, 1, 147
    Janka, H. T. 2012, ARNPS, 62, 407
    Branch, D., & Wheeler, J. C. 2017, Supernova Explosions (Berlin: Springer Nature)

Hirata, K., et al. 1987, PhRvL, 58, 1490; see also
    Bionta, R. M., et al. 1987, PhRvL, 58, 1494

Hoyle, F., & Fowler, W. A. 1960, ApJ, 132, 565

https://en.wikipedia.org/wiki/Supernova

https://en.wikipedia.org/wiki/Californium

https://en.wikipedia.org/wiki/List_of_supernova_remnants

Janka, H. T. 2012, ARNPS, 62, 407

Khokhlov, A., Müller, E., & Höflich, P. 1993, A&A, 270, 223; see also
    Hillebrandt, W., & Niemeyer, J. C. 2000, ARA&A, 38, 191

Leventhal, M., & McCall, S. L. 1975, Natur, 255, 690; see also
    Colgate, S. A., Petschek, A. G., & Kriese, J. T. 1980, ApJ, 237, L81

Livingstone, A. 2013, The Toronto Star, Tuesday, 1 January

Marsden, B. G. 1987a, IAU Circ, 4316, 24 Feb

Marsden, B. G. 1987b, IAU Circ, 4317, 25 Feb

Marsden, B. G. 1987c, IAU Circ, 4330, 4 Mar

Mayall, N. U. 1937, PASP, 49, 101; see also
    Schaeffer, R., et al. 1987, A&A, 184, L1
    Tyson, J. A., & Boeshaar, P. C. 1987, PASP, 99, 905

https://en.wikipedia.org/wiki/History_of_supernova_observation; see also
    Mayall, N. U. 1937, PASP, 49, 101
    Schaeffer, R., et al. 1987, A&A, 184, L1
    Tyson, J.A., Boeshaar, P.C., et al. 1987, PASP, 99, 905
    Baade, W. 1938, ApJ, 88, 285

Mihalas, D. 1963, PASP, 75, 256

Minkowski, R. 1941, PASP, 53, 224

Page, D., et al. 2020, ApJ, 898, 125

Raskin, C., et al. 2012, ApJ, 746, 62

Sat, K., & Suzuki, H. 1987, PhLB, 196, 267

Soderberg, A. M., et al. 2008, Natur, 453, 469

Starrfield, S., et al. 2004, ApJ, 612, L53

Suntzeff, N. B., & Bouchet, P. 1990, AJ, 99, 650; see also
    Suntzeff, N. B., et al. 1992, ApJ, 384, L33

Warner, B. 1974, MNRAS, 167, 61P

Webbink, R. F. 1984, ApJ, 277, 355
    Iben, I. Jr., & Tutukov, A. V. 1984, ApJS, 54, 335

Whelan, J., & Iben, I. Jr. 1973, ApJ, 186, 1997

Woosley, S. E., & Janka, H. T. 2005, NatPh, 1, 147

Chapter 17

Astrophysical Alchemy: Origins of the Chemical Elements

As Carl Sagan (1934–1996) once famously observed, "We are made of starstuff" (Sagan 1980). He meant, of course, that except for hydrogen and helium and a few other light elements, all of the other elements in our bodies—and indeed all the other elements and isotopes in the entire universe—were forged in the thermonuclear fires inside the stars and then redistributed into space by stellar winds or cataclysmic stellar explosions. How we came to understand this is the subject of the present chapter.

During the early decades of the 20th century, careful investigations of the spectra of the Sun and stars enabled astronomers to determine the relative abundances of the chemical elements in these objects. At the same time, other scientists were at work determining the chemical compositions of terrestrial rocks and of meteorites collected at the Earth's surface. By the late 1930s, it had become clear that the relative abundance distributions in all these disparate objects followed a surprisingly similar pattern (Figure 17.1; Anders & Grevesse 1989), which is sometimes called the "cosmic abundance curve."

There are of course differences. The Sun and stars contain far more hydrogen (H) and helium (He) than do terrestrial rocks or meteorites—or the lunar samples brought back from the Moon by the Apollo astronauts. Nevertheless, if the abundance distributions are matched at some particular element that two samples have in common, the remaining common elements exhibit strikingly similar patterns: the most abundant elements are carbon, nitrogen, and oxygen—collectively termed the "CNO elements." Then come the "iron-peak" elements: iron (Fe), cobalt (Co), and nickel (Ni). The light elements lithium (Li), beryllium (Be), and boron (B) have markedly lower abundances than any of the other light elements. And as one proceeds to higher atomic weights beyond the iron peak, the abundances exhibit a gradual decrease punctuated by pairs of abundance peaks.

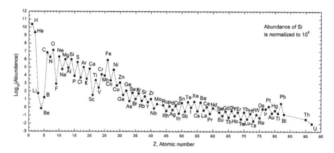

Figure 17.1. The logarithms of the abundances of the elements in the solar system. The vertical scale is arbitrarily normalized to an abundance of 10^6 atoms of silicon. The horizontal scale gives the atomic number, and each element is labeled with its chemical symbol. Source: commons.wikimedia.org/wiki/File: SolarSystemAbundances.png.

These surprising congruities called for explanations, and scientists took up the challenge. A very readable summary of the development of our current understanding of the processes of nucleogenesis has been provided by Jennifer A. Johnson (Johnson 2019).

17.1 Big Bang Nucleosynthesis

Following the discovery of the expansion of the universe (Hubble 1929), scientists came to understand that the very early universe had to have been extraordinarily dense and hot. As nuclear physics began to develop after World War II, they subsequently realized that nuclear-fusion reactions would be capable of building up heavier elements from primordial protons and neutrons (Gamow 1946). This idea was championed in particular by George Gamow and some of his colleagues.

However, the expansion of the early universe was so rapid that only about a half hour after the Big Bang, the temperature and density had already fallen so low that nuclear fusion reactions had essentially stopped (Weinberg 1977; see Figure 17.2). In addition, nuclei with atomic masses $A = 5$ and $A = 8$ are unstable and rapidly decay back into less massive nuclei, strongly suppressing the formation of heavier nuclei[1]. The consequent difficulty of accounting for the formation of *all* the chemical elements by primordial nucleosynthesis led to a decline of interest in the production of elements in the early universe. As an alternative, Fred Hoyle championed the idea that all the chemical elements were produced in stars (Hoyle 1946), which led him into a continuing academic conflict with Gamow (Halpern 2021). However, Hoyle's theory also had a significant drawback: where did the original hydrogen come from?

The lack of interest in primordial nucleosynthesis lasted until 1965, when Bell Laboratories scientists Arno Penzias and Robert Wilson discovered the cosmic

---

[1] The addition of a neutron or proton to the very stable nucleus ^{4}He produces ^{5}He or ^{5}Li, respectively, which immediately decay on timescales less than 10^{-21} seconds. Similarly, the fusion of two ^{4}He nuclei produces the isotope ^{8}Be, which immediately disintegrates back into two ^{4}He nuclei on a timescale of less than 10^{-16} seconds.

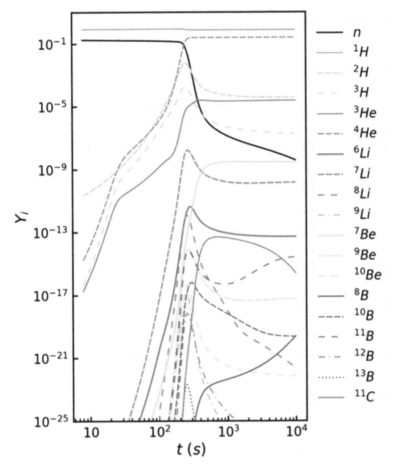

Figure 17.2. The relative abundances of the elements produced by Big Bang nucleosynthesis during the first several minutes of the expansion of the universe. The vertical axis gives the logarithm of the mass fraction of each species (with all the mass fractions adding up to unity), and the horizontal axis gives the time in seconds (on a logarithmic scale) since the start of the expansion of the universe. The different-colored curves represent the different species of particles, as listed at the right of the figure, with n = neutrons, D = ^{2}H = deuterium, ^{3}H = tritium, etc. Source: commons.wikimedia.org/wiki/File:Big_Bang_Nucleosynthesis_solved_using_PRIMAT_%287%29.png.

microwave background radiation (Penzias & Wilson 1965). This cooled and redshifted radiation remaining from the primordial cosmic fireball unequivocally demonstrated that our universe had originated in a "Big Bang." The following year, Princeton University cosmologist P. J. E. (Jim) Peebles published a calculation (Peebles 1966) of the abundances of ^{3}He, ^{4}He, and deuterium (^{2}H = D) produced in the early universe, assuming the temperature of the primordial fireball to have fallen to 3 K today. Caltech's Robert V. Wagoner and William A. Fowler, together with Fred Hoyle at Cambridge University, carried out detailed calculations of "Big Bang nucleosynthesis" (BBN) in 1967. As they put it, "If the recently measured microwave background radiation is due to primeval photons, then significant quantities of only D, ^{3}He, ^{4}He, and ^{7}Li can be produced in the universal fireball" (Wagoner *et al.* 1967).

This period has been well-described by Nobel laureate Steven Weinberg (1933–2021) in his very readable book, *The First Three Minutes.* Beginning at unimaginably high temperatures and densities, Weinberg describes how the primordial quark–gluon plasma subsequently combined into H and He in the proportions currently observed (approximately 75% H by mass and 25% He).

The current cosmological models that best agree with observations are called "Lambda Cold Dark Matter," or "ΛCDM," models. The Greek letter Λ represents the effects in Einstein's general-relativity equations of the mysterious "dark energy"—the existence of which was discovered only in the late 1990s—while "cold dark matter" (CDM) means unseen mass without any appreciable thermal energy. Hints had been around since the 1930s that the universe contains some form of matter that is not detectable through electromagnetic interactions. In the 1970s, careful measurements (Rubin & Ford 1970) by astronomers Vera Rubin (1928–2016) and W. Kent Ford showed that the rotation of distant galaxies required a stronger gravitational force—and thus more mass—than could be accounted for by the visible mass in stars, gas, and dust. This unseen mass has come to be called "dark matter." Its nature, too, is still a complete mystery.

In the early decades of the 21st century, two spacecraft—the Wilkinson Microwave Anisotropy Probe (WMAP) and most recently the *Planck* satellite—measured the properties of the cosmic microwave background (CMB) radiation with high precision. In the process, they determined highly accurate values for the six parameters that characterize the ΛCDM models. Results obtained in 2015 from the *Planck* spacecraft (Planck Collaboration: Ade *et al.* 2015), for example, yielded a value for the Hubble constant[2] of $H_0 = (67.8 \pm 0.9)$ km s^{-1} Mpc^{-1}. The reciprocal of the Hubble constant gives the age of the universe, which comes out to be 13.8 billion years. In addition, the *Planck* scientists were able to make accurate determinations of all six cosmological parameters. With the *Planck* values for these parameters, the ΛCDM cosmological models fit all of the constraints provided by the observational data to an accuracy of a few percent or better.

By 2016, analyses of data from the WMAP and *Planck* satellites also had produced similarly accurate determinations of the abundances of the light elements produced in the Big Bang (Cyburt *et al.* 2016). A comparison between the BBN predictions and observational determinations is shown in Table 17.1.

The quantity $Y_p = 0.24709 \pm 0.00025$ is the primordial mass fraction of helium (the fraction of a gram of primordial matter that consists of ^{4}He). The predictions for the primordial abundances of ^{4}He and deuterium (D = ^{2}H) from Big Bang nucleosynthesis are in excellent agreement with actual astronomical measurements. This is *not* the case for ^{7}Li, however. The sharp discrepancy between the observed primordial lithium abundance and the predictions is termed the "Lithium Problem."

---

[2] The Hubble constant H_0 measures the rate of expansion of the universe. For $H_0 = 68$ km s^{-1} Mpc^{-1}, galaxies at a distance of 1 Mpc (1 million parsecs, or about 3.26 million light-years) are streaming away from our location in the universe at a speed of 68 km s^{-1}; at 10 Mpc, the recession velocity is 680 km s^{-1}; and so on.

Table 17.1. BBN Element Abundances

	BBN Predictions	Observations
Y_p	0.2470	0.2449 ± 0.0040
D/H	2.579×10^{-5}	$(2.53 \pm 0.04) \times 10^{-5}$
^{3}He/H	0.996×10^{-5}	
^{7}Li/H	4.648×10^{-10}	$(1.6 \pm 0.3) \times 10^{-10}$
^{6}Li/H	1.288×10^{-14}	

While the Lithium Problem is a topic of ongoing research, the agreement between the predicted and observed abundances of the other light elements means that we actually do have adequate knowledge of the composition of the matter from which the first stars formed. We also know that all the remaining isotopes and elements had to have been produced inside the stars.

17.2 The Synthesis of the Elements in Stars

As we have seen in previous chapters, the thermonuclear fusion that takes place in the deep interiors of the stars not only generates the energy that powers the starlight, but also it transmutes the chemical elements initially present into others of larger atomic masses. Hans Bethe's 1939 calculations showed that H-fusion reactions like those that power the Sun end up producing ^{4}He, and in 1952 Ed Salpeter found that the 3α reaction can form ^{12}C directly from ^{4}He, bypassing the unstable ^{8}Be nucleus. Remarkably—by arguing that the relatively large amount of carbon present in the universe shows that the 3α reaction must proceed efficiently—Fred Hoyle actually predicted the existence of an energy level within the ^{12}C nucleus that enhances this rate before it was confirmed experimentally (Halpern 2021).

In 1957, two groups of scientists independently showed that—starting from the simplest elements, H and He—only a few types of nuclear-reaction processes are needed to produce all the remaining chemical elements and that these processes are likely to occur naturally during the later evolution of the stars. In that year, the British husband-and-wife astronomy team of E. Margaret Burbidge (1919–2020) and Geoffrey R. Burbidge (1925–2010), together with Caltech nuclear astrophysicist William A. Fowler (1911–1995) and the prolific British astrophysicist Fred Hoyle (1915–1995)—knighted in 1972 to become Sir Fred Hoyle—published a landmark scientific paper on stellar nucleosynthesis (Burbidge et al. 1957) familiarly known as "B^{2}FH" (see Figure 17.3), and Canadian nuclear astrophysicist Alastair G. W. Cameron (1925–2005) published a comparable but less detailed paper (Cameron 1957). At the time, the main sites for stellar nucleogenesis—and the subsequent distribution of the newly created elements into space—seemed likely to be supernova explosions. While supernovae are still thought to be prime nucleosynthetic sites, the intervening half century and more have revealed a more nuanced picture.

Figure 17.3. The iconic "B^{2}FH" team in 1971. From left to right: E. Margaret Burbidge, Geoffrey R. Burbidge, William A. Fowler, and Fred Hoyle. They are admiring a model locomotive given to Fowler as a birthday present. Credit: AIP Emilio Segrè Visual Archives, Clayton Collection; copyright American Institute of Physics.

The first stage of stellar nucleosynthesis consists of H burning in the cores of Main Sequence stars. As we have seen in previous chapters, this stage transmutes H in the deep, nuclear-burning core of a Main Sequence star into He. In the course of processing through the various H-burning nuclear reactions, this stage also produces some isotopes of C, N, O, F, Ne, and Na.

The next stage consists of He burning through the 3α reaction in the cores of red-giant stars at temperatures of 1–2×10^8 K and densities of some 10^5 g cm^{-3}. These reactions produce ^{12}C, ^{16}O, ^{20}Ne, and ^{24}Mg.

As we saw in the previous chapter, at still higher temperatures of some 10^9 K, photon energies are high enough to photodisintegrate some of the ^{20}Ne into ^{16}O plus α particles. The newly created α particles are captured successively by other ambient nuclei to produce ^{24}Mg, ^{28}Si, ^{32}S, ^{36}A, ^{40}Ca, and perhaps ^{44}Ca and ^{48}Ti. Burbidge *et al.* named this the "α-process."

At temperatures above about 3×10^9 K, many different nuclear processes occur in a kind of statistical equilibrium, and Burbidge *et al.* named this the "*e*-process." It is responsible for synthesizing the "iron-peak" elements vanadium (V), chromium (Cr), manganese (Mn), iron (Fe), cobalt (Co), and nickel (Ni).

Up to this point, the synthesis of heavier elements is able to generate enough energy to support a star against the weight of the overlying layers. However, the binding energy of a nucleus reaches a maximum at iron, so nuclear reactions that combine any heavier elements represent an energy *sink* rather than an energy *source*.

The next process involved in stellar nucleosynthesis involves the production of neutrons and their absorption by atomic nuclei on timescales of hundreds to thousands of years. These timescales are long enough to allow any β-unstable nuclei to decay, so that neutron captures take place on the stable daughter nuclei. This slow process of

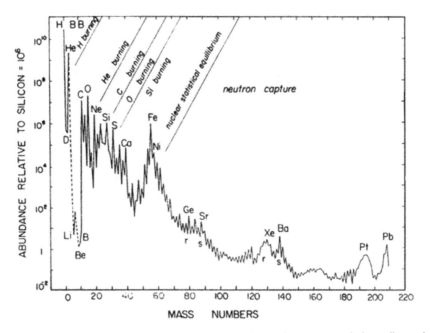

Figure 17.4. The relative abundances of the elements produced by various processes during stellar nucleosynthesis. The abundance of Si is arbitrarily set to 10^6. The various stages of stellar nucleosynthesis responsible for the production of different groups of elements are indicated on the diagonal lines. Thus, He burning produces ^{12}C, ^{16}O, ^{20}Ne, and ^{24}Mg; "nuclear statistical equilibrium" produces the iron-peak elements, and so on. The label "s" under Sr and Ba denotes the *s*-process as the source of these nuclei, while the "r" below Ge and Xe denotes the *r*-process. Source: From Pagel, B. E. J. 1997, *Nucleosynthesis and Chemical Evolution* (New York: Cambridge Univ. Press); https://ned.ipac.caltech.edu/level5/Pagel/figure1_4.jpg; used under license 79487 from PLSClear.

neutron capture—called the *s*-process—is responsible for producing most of the isotopes in the range of atomic weights $25 \leqslant A \leqslant 70$ and for many more isotopes in the range $70 \leqslant A \leqslant 204$ (Truran & Iben 1977). It also produces peaks in the element abundances at $A = 90$, 138, and 208 (see Figure 17.4).

The source of neutrons for the *s*-process comes from further H burning of the products of pure He burning. This occurs in the stage of stellar evolution beyond core He burning, when the H- and He-burning shells overlying the degenerate C/O core of a star undergo sequential thermal pulses (TPs) on the asymptotic giant branch (AGB). In such a thermally pulsing AGB star, the surface convection zone periodically extends deep into the stellar interior, dredging up the products of core He burning and mixing them into the H-burning shell (Iben 1975).

For AGB stars with masses $M < 2\ M_{Sun}$, during the ignition of the H-burning shell in a thermal pulse, the reaction sequence[3] ^{12}C(p,γ)^{13}N(β$^+$ν)^{13}C produces a thin pocket of ^{13}C in the He- and C-rich mantle. Neutrons are subsequently generated from the ^{13}C by the reaction

---

[3] [The expression ^{12}C(p,γ)^{13}N is a compact way of writing the reaction ^{12}C + p → ^{13}N + γ].

$$^{13}C + \alpha \rightarrow {}^{16}O + n$$

at a temperature of about 0.9×10^8 K (Käppeler et al. 2011).

For the more massive AGB stars in the range $2\,M_{Sun} < M < 8\,M_{Sun}$, H burning in a thermal pulse occurs via the CNO cycle, which converts the CNO nuclei mainly to ^{14}N. During the quiescent interpulse phase, the ^{14}N is converted into ^{22}Ne by the reaction sequence $^{14}N(\alpha,\gamma)^{18}F(\beta^+\nu)^{18}O(\alpha,\gamma)^{22}Ne$, and during the next He-burning pulse at about 3×10^8 K, further alpha captures generate neutrons from the reaction (Iben 1975; Käppeler et al. 2011)

$$^{22}Ne + \alpha \rightarrow {}^{25}Mg + n.$$

This process occurs at the high-temperature base of the thin convective shell near the peak of a thermal pulse that takes an intermediate-mass star up the AGB. The maximum extent of the convective shell formed during a thermal pulse (TP) extends from the region with the highest He-burning rate up to just below the H/He boundary.

After the temporary TP-induced convective shell disappears, the base of the outer convection envelope extends downward into the star during the so-called "third dredge-up" (TDU) phase and brings newly synthesized elements up to the surface. This enriches the surface layers in ^4He, ^{22}Ne, ^{25}Mg, ^{56}Fe, and the s-process elements. Because AGB stars less massive than $8\,M_{Sun}$ ultimately become white dwarfs, they lose much of their mass through strong stellar winds, which of course carry the newly synthesized elements out into the interstellar medium. In the atomic-mass range $70 \leqslant A \leqslant 204$, the resulting s-process abundances fit the observed solar system abundances very well (Truran & Iben 1977). Most of these nuclei are in fact produced by intermediate-mass stars ($2\,M_{Sun} \leqslant M \leqslant 8\,M_{Sun}$). These stars also produce appreciable amounts of elements with atomic masses $A < 70$, although most of these lighter heavy elements are produced in supernova explosions of stars with masses $M \geqslant 15\,M_{Sun}$.

In contrast to the s-process, another nucleosynthetic process—the r-process—involves neutron captures on very rapid timescales of 10 seconds or less, significantly faster than the rates of the intervening β-decays. This process synthesizes numerous elements in the range $70 \leqslant A \leqslant 209$, including uranium (U) and thorium (Th) and other transuranic elements. It may also be responsible for synthesizing some isotopes of Ca and Ti, and it produces abundance peaks at atomic weights $A = 80$, 130, and 194. In the r-process, a large neutron flux adds neutrons rapidly to iron-peak elements.

The primary site of r-process nucleosynthesis—long thought to be produced in supernova explosions—was discovered only in 2017, when the Laser Interferometer Gravitational-Wave Observatory (LIGO) and its sister observatory Virgo both detected a gravitational-wave event (Cho 2017)—GW170817[4]—that matched the-

---

[4] This gravitational-wave event was detected on 2017 August 17.

oretical calculations for the inspiral and collision of a pair of neutron stars. Less than 2 seconds after the gravitational-wave signal ended, the *Fermi* INTEGRAL gamma-ray telescope detected a flash of gamma rays. And 11 hours later, optical and infrared telescopes observed a "kilonova" glowing brightly from newly synthesized radioactive heavy elements. This event was subsequently detected as a radio and x-ray source two weeks after the gravitational-wave outburst. Comparison of ultraviolet–optical–infrared observations from 24 telescopes on seven continents enabled scientists to construct the light curve from the source (Kashiwal *et al.* 2017). Comparisons with theoretical models of neutron–star mergers enabled them to conclude that 0.05 M_{Sun} of elements substantially heavier than those produced in supernovae had been created in the event. The radioactive decay of these heavy elements powered the light curve of the kilonova. The long-sought site of *r*-process nucleosynthesis finally had been found! If neutron–star mergers do dominate *r*-process heavy-element production, the authors conclude that an event like GW170817/SSS17a[5] must occur in our Galaxy every 20,000 to 80,000 years (Driut *et al.* 2017). The debris resulting from the explosive disruptions of pairs of neutron stars distributes the *r*-process elements produced in the collision into the interstellar medium.

The final process discussed by Burbidge *et al.* is called the "*p*-process." This process is responsible for the production of some 35 proton-rich nuclei ("*p*-nuclei") that cannot be created by either the *s*-process or the *r*-process. The abundances of the *p*-nuclei are typically ten to a thousand times less than those of the *s*- or *r*-process nuclei. Most of these isotopes have even values of the atomic number A. The astrophysical conditions under which the *p*-process occurs have been controversial. In their pioneering 1957 papers, Cameron and Burbidge *et al.* suggested that the *p*-nuclei were produced in H-rich envelopes of supernovae of Type II (SN II) by (p,γ) and (γ,n) reactions on *r*- and *s*-process "seed" nuclei or in the outer parts of the envelope of a supernova of Type I (Woosley & Howard 1978). Two decades later, several groups of investigators suggested that most of the *p*-nuclei in the solar system were produced by the photodisintegration of *s*-process seed nuclei in SN II (Audouze *et al.* 1978). In 2011, Travaglio *et al.* argued that most *p*-nuclei were created by sequences of photodisintegration and β decays in both core-collapse supernovae (SN II) and in supernovae of Type Ia (SN Ia; Travaglio *et al.* 2011). In SN II, they found that the explosive burning of Ne and O does produce *p*-nuclei but that nuclides with A < 120 are underproduced. In contrast, they found that SN Ia produce large amounts of both light and heavy *p*-nuclei.

While the nucleosynthesis of the elements in stars continues to be a subject of active research, there is no doubt that the majority of the elements and their isotopes in the universe have been produced inside successive generations of stars. The periodic table of the elements shown in Figure 17.5 provides a graphic illustration of the respective contributions of various astronomical sources to the creation of the different chemical elements.

---

[5] SSS denotes a discovery from the Swope Supernova Survey carried out at the Swope Telescope at Las Campanas Observatory in Chile.

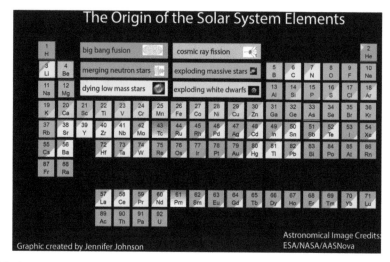

Figure 17.5. A summary illustration of the contributions of various astronomical sources to the production of the chemical elements. Source: Jennifer A. Johnson/The Ohio State University; NASA; ESA.

References

Anders, E., & Grevesse, N. 1989, GeCoA, 53, 197; see also
 Asplund, M., et al. 2009, ARA&A, 47, 481
 Grevesse, N., et al. 2015, A&A, 573, A27
Audouze, J., & Truran, J. W. 1975, ApJ, 202, 204; see also
 Arnould, M. 1976, A&A, 46, 117
 Woosley,, & Howard, 1978, ApJS, 36, 285
Burbidge, E. M., Burbidge, G. R., Fowler, W. A., & Hoyle, F. 1957, RvMP, 29, 547
Cameron, A. G. W. 1957, PASP, 69, 201
Cho, A. 2017, Sci., 358, 1520; see also
 Grant, A. 2017, PhT, 70, 19
Cyburt, R. H., Fields, B. D., Olive, K. A., & Yeh, T. H. 2016, RvMP, 88, 015004
Driut, M. R., et al. 2017, Sci., 358, 1571
Gamow, G. 1946, PhRv, 70, 572; see also
 Alpher, R. A., Bethe, H., & Gamow, G. 1948, PhRv, 73, 803
Halpern, P. 2021, Flashes of Creation (New York: Basic Books)
Hoyle, F. 1946, MNRAS, 106, 343
Hubble, E. 1929, PNAS, 15, 168
Iben, I. Jr. 1975, ApJ, 196, 525
Johnson, J. A. 2019, Sci., 363, 474
Käppeler, G., Gallino, R., Bisterzo, S., & Aoki, W. 2011, RvMP, 83, 157
Kashiwal, M. M., et al. 2017, Sci., 358, 1559
Peebles, P. J. E. 1966, ApJ, 146, 542
Penzias, A. A., & Wilson, R. W. 1965, ApJ, 142, 419
Planck Collaboration: Ade, P. A. R., et al. 2015, arXiv:1502.01589
Rubin, V. C., & Ford, W. K. 1970, ApJ, 159, 379; see also
 Rubin, V. C., Ford, W. K. Jr., & Thonnard, N. 1987, ApJ, 238, 471

Yaeger, A. J. 2021, Bright Galaxies, Dark Matter, and Beyond: The Life of Astronomer Vera Rubin (Cambridge, MA: MIT Press)

Sagan, C. 1980, Cosmos (New York: Random House), 233

Travaglio, C., Röpke, F. K., Gallino, R., & Hillebrandt, W. 2011, ApJ, 739, 93

Truran, J. W., & Iben, I. Jr. 1977, ApJ, 216, 797

Wagoner, R. V., Fowler, W. A., & Hoyle, F. 1967, ApJ, 148, 3

Weinberg, S. 1977, The First Three Minutes (New York: Basic Books), 112; see also Wagoner, R. V. 1973, ApJ, 179, 343

Woosley, S. E., & Howard, W. M. 1978, ApJS, 36, 285

Chapter 18

Neutron Stars

In 1932, British physicist James Chadwick (1891–1974), working at the Cavendish Laboratory at Cambridge University, discovered the neutron, a fundamental particle slightly more massive than the proton, but with no electric charge (Chadwick 1932). Like the electron and proton, the neutron has half-integral spin; that is, it carries an intrinsic angular momentum of ½ℏ, where ℏ is Planck's constant divided by 2π. The neutron and proton interact by means of a very short-range force called the "strong nuclear force," which binds them together stably in atomic nuclei. In contrast, an isolated neutron spontaneously decays in about 12 minutes into a positively charged proton, a negatively charged electron, and a ghostly neutrino. Because they have no electric charge, neutrons essentially do not interact with electrons.

Like protons and electrons, neutrons obey Fermi–Dirac quantum statistics and are called "fermions." As we saw in Chapter 15, Chandrasekhar showed in 1931 that theoretical models consisting of fermions with masses equal to that of the electron and which are in hydrostatic equilibrium at zero temperature have masses and radii consistent with those of the white dwarf stars. Statistical physics tells us that the pressure of a degenerate gas of non-relativistic fermions is inversely proportional to the mass of the fermion. Other things being equal, the pressure support provided by a degenerate neutron gas is consequently smaller than that provided by a degenerate electron gas by the ratio of the neutron mass to the electron mass, $m_n/m_e = 1838.65$. A stellar mass of degenerate neutrons must therefore contract to a much higher density than that of a white dwarf in order for the pressure provided by the degenerate neutrons to become large enough to support the star against its own self-gravity. Instead of the planetary dimensions of a white dwarf—several thousand kilometers—the radius of a neutron star is only of the order of 10 km—about the size of a city. Of course, such small dimensions also mean that the average density of a neutron star is around 5×10^{14} g cm^{-3}—greater than the density of a typical atomic nucleus!—rather than the "mere" 10^6 g cm^{-3} average density of a white dwarf. That is, while a chunk of white dwarf matter about the size of a sugar cube

doi:10.1088/2514-3433/acce33ch18

would weigh as much as a small car, a similar-sized chunk of neutron-star matter would weigh about one hundred times as much as the Great Pyramid of Giza! The high density of a neutron star also means that the surface gravity is some five billion times as large as the surface gravity of the Sun. The gravitational potential energy of a particle at the surface of a neutron star is about a third of its rest mass, which means that Einstein's general relativistic theory of gravitation must be used to construct models of neutron stars, rather than Newton's non-relativistic theory.

In 1939, American physicist Richard C. Tolman (1881–1948) provided a mathematical foundation for the construction of general relativistic stellar models, and J. Robert Oppenheimer (1904–1967) and his student George M. Volkoff (1914–2000) utilized it to construct the first mathematical models of neutron stars (Tolman 1939; Oppenheimer & Volkoff 1939), each idealized model consisting of a cold Fermi gas of non-interacting neutrons. These early calculations showed that the pressure exerted by an ideal neutron gas (i.e., neglecting the strong interactions that bind atomic nuclei together) cannot support an object with a mass greater than about three-quarters of the mass of the Sun. Although that value for the limiting mass of a neutron star—called the Tolman–Oppenheimer–Volkoff (TOV) limit, which is analogous to the Chandrasekhar limiting mass for white dwarfs—has since been increased, the existence of such a limiting mass is not in doubt. However, without observational evidence for their existence at the time, neutron stars remained of interest primarily to theorists for the next several decades, even though various ideas were suggested as possible ways to discover them.

Proof that neutron stars really do exist came from a completely unexpected direction, as happens so often in astronomy. In the late 1960s, graduate student Jocelyn Bell (Figure 18.1) was working on a radio astronomy research project at Cambridge University in England under the direction of her Ph.D. thesis advisor, Antony Hewish (1924–2021). She discovered a small number of celestial radio sources that were emitting pulses with very precise, short periods—a few seconds or less. Initially thinking the pulses were being produced by some form of terrestrial interference, Bell and Hewish checked carefully before concluding that the signals really did come from some extraterrestrial source. Facetiously dubbing them LGM1, LGM2, etc.—for "Little Green Man 1, 2," etc., as if they were from some extraterrestrial civilization—Bell and Hewish carefully investigated their properties before publishing their astonishing discovery in 1968 (Hewish et al. 1968). Hewish shared the 1974 Nobel Prize in Physics for this discovery, but Bell's contribution was ignored by the Nobel committee. Many scientists regarded this as inappropriate. Since then, however, Dame Susan Jocelyn Bell Burnell has been honored with numerous awards, and in 2018 she was awarded the Special Breakthrough Prize in Fundamental Physics—currently the most lucrative academic prize in the world—for her role in this work.

Immediately after the discovery of these mysterious pulsating radio sources, other astronomers jumped in, with many of the largest radio telescopes around the world becoming involved. The 250-foot Lovell radio telescope at the Jodrell Bank Observatory in the United Kingdom, the 210-foot-diameter Parkes telescope in Australia, the 300-foot dish at the Green Bank Observatory in West Virginia, and

Figure 18.1. Susan Jocelyn Bell (Burnell). Credit Roger W. Haworth; Source: commons.wikimedia.org/wiki/File:Susan_Jocelyn_Bell_(Burnell),_1967.jpg.

the 1000-foot (305-meter) telescope at Arecibo, Puerto Rico[1] were among the large radio telescopes that became actively involved in utilizing their large collecting areas to search for the signals from these faint radio sources. Could other pulsating radio sources be found, astronomers wondered? Many soon were, and astronomers named them "pulsars." Could they be detected at optical wavelengths? For some time, optical searches came up empty. Then in 1968, David H. Staelin (1938–2011) and his collaborators discovered the existence of a pulsar with a period of about 33 milliseconds (Staelin *et al.* 1969) located within the Crab Nebula, the remnant of a supernova that had exploded in 1054 AD (see Figure 16.4). John Cocke (1937–2022) and his colleagues at the University of Arizona's Steward Observatory quickly set their sights on this object, and in 1969 they discovered that a very blue object near the center of the Crab was undergoing optical pulsations with the very same period as the radio pulsar (Cocke *et al.* 1969). This result was immediately confirmed by Ed Nather (1926–2014) and his colleagues at the University of Texas (Nather *et al.* 1969).

---

[1] Both the 300-foot Green Bank radio telescope and the 1000-foot Arecibo telescope have since collapsed, the former in 1988 and the latter in 2020. The 300-foot dish was replaced in 2001 by the fully steerable 100-meter Robert C. Byrd Green Bank Telescope, while the future of the Arecibo Observatory is still under discussion at the time of this writing.

The Crab pulsar provided the clue that led Cornell University astrophysicist Tommy Gold (1920–2004) to identify pulsars as rapidly rotating neutron stars. Gold reasoned that the extreme regularity of the pulsations required a flywheel of enormous mass, which could only be provided by a rotating star. And the extreme shortness of the rotation period of the Crab pulsar meant that the system had to be smaller than a white dwarf, or else the equatorial rotation speed would exceed the speed of light. A neutron star was the only candidate that fit the bill (Gold 1968). Although the concept of a rapidly rotating "lighthouse" beacon embedded in a flywheel provided by the mass of a rapidly rotating neutron star could explain the regular pulsations produced by a pulsar, what was the source of the lighthouse beam? Gold argued that an extremely strong magnetic field could produce the observed beaming.

Not long after pulsars had been discovered, observers also found that their rotation periods were slowly increasing (Davies *et al.* 1969). The rate of change of the period is very slow, however—typically some 10^{-15} s s^{-1}—corresponding to spindown ages of some 10 million years. Spindown occurs because the strong magnetic field of the pulsar, rotating at the rapid rate indicated by the regular pulses, generates copious electromagnetic radiation that slowly drains the rotational kinetic energy from the neutron star (Ostriker & Gunn 1969). Because the electromagnetic energy radiated by the rotating pulsar is proportional to the square of the magnetic field strength, measurements of the period and its gradual rate of change in time enabled astronomers to estimate the strength of the pulsar's magnetic field. The resulting estimates of the surface field strengths of typical pulsars range from about 20 billion gauss (2×10^{10} gauss, or some 40 billion times larger than the magnetic field strength at the surface of the Sun) to about 2×10^{13} gauss (Manchester & Taylor 1977). The lower value is roughly what would be obtained if a field initially equal to the surface magnetic field strength of the Sun were "frozen" into the core of a pre-supernova star—due to its high electrical conductivity—and then collapsed to the density of a neutron star. If the initial field were instead as strong as the field of a Main Sequence magnetic A star, the core collapse would instead generate fields a thousand times or more larger than this, which is consistent with the larger value given above.

Gold's former student, Peter Goldreich, and Goldreich's colleague William H. Julian developed a more detailed model of the actual magnetosphere of a strongly magnetized, rapidly rotating neutron star (Goldreich & Julian 1969; Ostriker & Gunn 1969). They pointed out that the strong, rapidly rotating magnetic field also generates an extremely strong electric field that points outward (or inward) along the magnetic field lines. The field is strong enough to yank electrons (or positrons) out of the pulsar's polar caps, rapidly accelerate them to relativistic energies, and produce an electron–positron pair-production cascade that beams energy along the magnetic field lines. While this general picture of the pulsar emission mechanism was proposed more than half a century ago, the detailed processes by which the strongly beamed electromagnetic radiation is produced still remain a subject of active investigation.

The discovery of the pulsar in the Crab Nebula not only proved that neutron stars really do exist, but also it showed that they are born in supernova explosions, as originally suggested by Baade and Zwicky in 1934 (Baade & Zwicky 1934), and that they come into existence with extraordinarily strong magnetic fields and spinning very rapidly. By hindsight, none of this should have been a surprise. As we saw in Chapter 16, one of the models for a supernova explosion involves the collapse of an iron core inside a massive star when it grows too massive to be supported any longer by degenerate-electron pressure. If the core collapse can be halted by degenerate-neutron pressure, while the rest of the massive star explodes back out into space, the result is a newly born neutron star in the midst of a supernova remnant. Using conservation of angular momentum during the core collapse to scale the rotation period from that of the Sun yields a rotation period less than a millisecond for the newly formed neutron star. Although the first pulsars to be discovered had periods ranging from a fraction of a second (like the Crab pulsar) to a few seconds, the first millisecond pulsar was actually discovered in 1982 (Backer *et al.* 1982). Named PSR B1937+21, it had a rotation period of 1.557708 ms. As of this writing, more than 2000 pulsars have been discovered, with periods ranging from a few milliseconds (PSR B1937+21) (Backer *et al.* 1982) up to 23.5 seconds (PSR J0250+5854), the longest period found to date (Tan *et al.* 2018). The mean period of all pulsars is about 0.8 seconds.

As pointed out by Robert Duncan and Christopher Thompson (Duncan & Thompson 1992), the existence of millisecond pulsars indicates that it is possible for neutron stars to be formed with even stronger magnetic fields than those of typical pulsars. They argued that the coexistence of rapid and strong differential rotation with the chaotic turbulence in the collapsing core of a supernova provide conditions for efficient dynamo action—like the dynamo activity inside the Sun, but on a very much stronger scale. Their calculations showed that fields as high as 10^{14}–10^{15} gauss —up to 100 times stronger than the fields in typical pulsars—can be produced in this way. Because the fields are so strong and the neutron stars are rotating so rapidly, these so-called "magnetars" lose their rotational energy very quickly. Twenty-four magnetars had been discovered as of July 2021 (https://en.wikipedia.org/wiki/Magnetar).

Duncan and Thompson also conjectured that starquakes (see below) can perturb the enormously strong magnetic field of a magnetar, producing powerful bursts of soft gamma rays, such as those emitted on 1979 March 5 by a source within supernova remnant N49 in the Large Magellanic Cloud (Cline *et al.* 1982). This hypothesis has since become the generally accepted explanation for "soft gamma-ray repeaters" (SGRs). It has also been suggested that magnetars may be the sources of the Fast Radio Bursts (FRBs) discovered in 2007 (Lorimer *et al.* 2007). Observations of FRB 200428, detected in April 2020 at several wavelengths, provide compelling evidence for its association with the magnetar SGR 1935+2154 (Clary 2020; Berkowitz 2021).

18.1 Inside a Neutron Star

So, what do we know about the interior of a neutron star? (See Figure 18.2.) First, it does not consist solely of neutrons all the way from the center to the surface. Calculations of the equilibrium composition of matter in zero-temperature neutron-

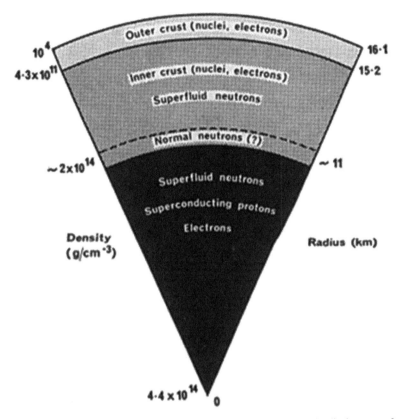

Figure 18.2. Schematic illustration of the internal structure of a neutron star. Credit: heasarc.gsfc.nasa.gov/docs/objects/binaries/neutron_star_structure.html; image courtesy of NASA.

star models (Shapiro & Teukolsky 1983) show that at densities less than about 8×10^6 g cm^{-3}—roughly the central density of a 0.8 M_{Sun} C/O white dwarf—the surface layers of a neutron star consist of iron. In these outermost layers, the neutron-star temperature may be of the order of 10^7 K. At these temperatures, matter sufficiently close to the surface is fluid rather than solid, but as we proceed inward the Coulomb interactions between the increasingly densely packed, high-Z nuclei cause them to freeze into a crystalline lattice—like the inner core of a cool white dwarf—which is called the "crust" of the neutron star. In addition, it becomes energetically favorable for electrons to be captured into the atomic nuclei at such high densities, making the nuclei increasingly neutron-rich as we go deeper into the neutron star. Ultimately, above a density of about 4.3×10^{11} g cm^{-3}, it becomes energetically favorable for free neutrons to appear within the lattice of neutron-rich nuclei; we have reached the point of "neutron drip." At such very high densities, the Fermi energies of the residual electrons, which are necessary for matter to remain overall electrically neutral, are so large that they prevent the neutrons from decaying, as they do in the laboratory. The neutrons thus remain stable throughout the rest of the neutron star interior. Consequently, the structure of matter in this inner part of the neutron star crust consists of a lattice of extremely neutron-rich nuclei immersed in a sea of relativistic, degenerate electrons and an increasingly dense and degenerate neutron fluid.

Conditions in the inner crust become even more extreme as we proceed still further inward, until we reach a density of about 2×10^{14} g cm^{-3}. At this point the remaining nuclei dissolve, leaving matter that consists primarily of neutrons, together with a smaller admixture of protons and electrons. This is approximately the density of typical atomic nuclei on Earth. In this sense, the interior of a neutron star thus resembles a gigantic atomic nucleus. Densities can be higher still in the deep interior of more massive neutron stars, but the properties of such extremely dense matter are largely speculative. Does a pion condensate develop? Is there a transition from a neutron fluid to a neutron solid? Or does some still-more-exotic state like quark matter come into existence?

As is the case for electron-degenerate white dwarfs, the radius of a neutron-degenerate star depends upon its mass. Unlike white dwarfs, however—for which a gas of non-interacting fermions (electrons) provides a good first approximation to the mass–radius relationship (see Figure 15.3)—for a neutron star the strong nuclear force has a very substantial effect on the internal structure and maximum mass. If two nucleons—neutrons or protons—are separated by more than a few times the diameter of an atomic nucleus[2], they feel essentially no force at all. However, when they are closer than this, they feel a very strong attraction that pulls them into contact and binds them tightly together. Within a fraction of this distance, however, the force becomes very strongly repulsive, so that the nucleons resemble very tiny hard spheres. In a sense—and on a completely different scale—the strong nuclear force resembles the forces between atoms that bind them into molecules.

It is difficult to determine accurately the strong nuclear force between isolated nucleons, although many nuclear-physics experiments since the mid-20th century have endeavored to accomplish this. The result is that several different parametrized versions of the strong nuclear force have been proposed. Each leads to a somewhat different structure and a somewhat different mass for the neutron star (see Figure 18.3). Each of the curves in this figure (Riley *et al.* 2021) corresponds to a relationship between the neutron-star mass (the vertical axis) and the corresponding radius (the horizontal axis) predicted by a particular model for the strong nuclear force. The letters in the legend are abbreviations for the last names of the authors of the particular model. (This is the neutron-star equivalent of Figure 15.3 for white dwarfs, but with the horizontal and vertical axes interchanged.) For our present purposes, the details of the various models are unimportant; what is important is the large spread in predicted radii for a given mass. In each case, the lighter curve is for a non-rotating model, while the heavier curve is for a neutron star rotating at 346 Hz. For pulsar J0740+6620, for which Fig. 18.3 summarizes the observations, the most probable mass and radius are about 2.1 M_{Sun} and 12.4 km. Although the uncertainty in the mass determination is only about 0.1 M_{Sun}, the radius could be as large as

---

[2] The diameter of an atomic nucleus is typically a few fermis—10^{-13} cm, 100,000 times smaller than the typical dimensions of an atom.

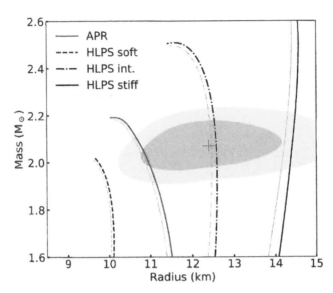

Figure 18.3. Mass–radius relations for various neutron-star equations of state compared with the observational determinations of the mass and radius of pulsar J0740+6620. Credit: Riley *et al.* (2021).

15 km (for the stiffest equation of state) or as small as about 10.5 km (for the softer equations of state). As of 2016, masses had been determined for about 35 neutron stars, ranging from 1.17 M_{Sun} to 2.0 M_{Sun}, with a peak at 1.33 M_{Sun} ± 0.1 M_{Sun}, and radii for more than a dozen were found to range from 9.9 km to 11.2 km (Özel & Freire 2016).

18.2 Superfluid Neutrons and Superconducting Protons

In 1959, well before the discovery of pulsars, Soviet physicist Arkady B. Migdal (1911–1991) suggested that the neutrons inside neutron stars might actually exist in a superfluid state (https://en.wikipedia.org/wiki/Arkady_Migdal#cite_ref_eduspb_3-0). After pulsars had been identified as rapidly rotating neutron stars, scientists immediately agreed that the neutrons in their interiors do exhibit superfluidity—that is, they flow without any apparent viscosity—while the protons are superconducting—that is, they carry an electrical current without experiencing any electrical resistance (Baym *et al.* 1969a). Discovered in a terrestrial laboratory in 1911, superconductivity was explained in 1957 by the so-called "BCS theory," (Bardeen *et al.* 1957) in which—at sufficiently low temperatures—charged fermions (electrons, in the original BCS theory) form "Cooper pairs" in a correlated quantum state that is bound too tightly to be broken by thermal fluctuations or by the random scattering processes that produce electrical resistance. Additional ultralow-temperature experiments in 1924 showed that the noble gas He, which becomes a liquid below about 4 K, undergoes another transition below 2.2 K after which it can flow through narrow tubes without apparent friction. Dubbed a "superfluid," under these conditions liquid He can be thought of as a mixture of two fluids, one of which is a normal liquid, while in the other fluid the atoms are condensed into a single quantum state that flows without experiencing viscous dissipation.

Although under terrestrial conditions matter becomes superfluid or superconducting only at very low temperatures, at the enormous densities inside neutron stars, the critical temperatures at which these transitions occur are far higher than the $\sim 10^8$ K internal temperatures estimated for the interiors of "mature" neutron stars (Haskell & Sedrakian 2018). The superfluid neutrons form an array of vortices —which carry the angular momentum of a spinning neutron star within the inner crust and the core—while the superconducting protons form an array of magnetic flux tubes, each carrying a quantum of magnetic flux, which enable the strong magnetic field of the neutron star to thread through the superconducting protons, as in a typical Type II superconductor on Earth. Both the superfluidity of the neutrons and the superconductivity of the protons can have significant effects on the dynamical behavior of a neutron stars, as we discuss next.

In 1969, Richard N. Manchester was at the Parkes Observatory in Australia helping Venkataraman Radhakrishnan (1929–2011) with observations of PSR B0833–45—a pulsar with a spin period of 89 milliseconds in a supernova remnant in the constellation Vela—when they observed a remarkable speedup in the pulsar rotation period (see Figure 18.4; Radhakrishnan & Manchester 1969; Manchester 2018). At the same time, Paul Reichley and George Downs were using the Goldstone tracking station to monitor the spindown of the same pulsar and observed the same event (Reichley & Downs 1969). Many more such "glitches" in the rate of spin—abrupt increases in the slow decline of the neutron-star rotation rate by a small amount, which then proceed to decline again, gradually approaching the original spindown rate over days or months or years—have since been observed in the Vela pulsar and in many other pulsars (Espinoza *et al.* 2011).

The Vela glitch was immediately interpreted by Gordon Baym and his colleagues in terms of a very simple and appealing model (Baym *et al.* 1969b). Because the neutron star is rotating rapidly, centrifugal force causes it to take the shape of an oblate spheroid (a slightly squashed sphere). As the pulsar spins down, the centrifugal force weakens, and stress builds up in the solid crust. When the stress becomes large enough, the crust fractures, producing a "starquake." In the process, the crust becomes slightly less oblate, the moment of inertia of the neutron star decreases slightly, and its rotation temporarily increases by a small amount. Baym *et al.* therefore considered the neutron star to consist of two major components: the crust and charged particles, which are strongly coupled together by the enormous magnetic field and effectively behave as a single unit, and the superfluid neutrons, which are very weakly coupled to the charged particles. While the former react promptly to the starquake, the change in rotation of the neutron star is communicated only gradually to the superfluid neutrons on a much longer timescale because of their very weak frictional coupling to the crust. According to Baym *et al.*, the post-glitch behavior of the Vela pulsar thus provides firm evidence that the interior of a neutron star does indeed contain superfluid neutrons.

Of course, the story is more complicated than this simple "crustquake" model, which cannot account for the wide variety of post-glitch behaviors that have since been observed. In 1972, physicist David Pines (1924–2018) and his colleagues proposed an alternative "corequake" model in which the superfluid neutron vortices

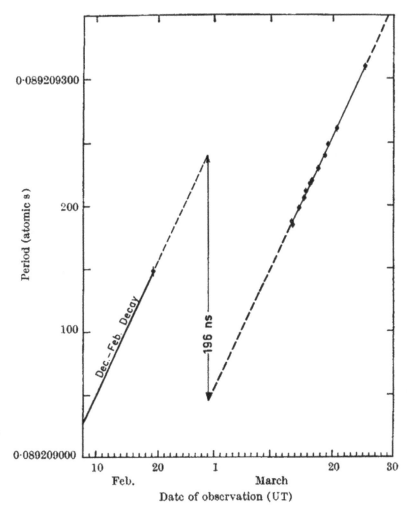

Figure 18.4. A glitch in the Vela pulsar, PSR B0833–45. The vertical axis gives the pulsar rotation period in seconds, while the horizontal axis gives the date of observation. Credit: Manchester, R. N. (2018), copyright International Astronomical Union.

that carry the angular momentum of the neutron-star core abruptly detach ("unpin") from the inner crust and then reattach ("repin"), transferring angular momentum to the crust and increasing the spin of the neutron star by a small amount. This and similar models have since been developed in considerable detail, some including "creep" of the neutron vortices (Alpar & Baykal 2006; Manchester 2018).

18.3 Neutron Star Binaries

Most normal stars exist in binary (or multiple) star systems. In a sense it is therefore not surprising that about 5% of all known neutron stars exist in binary systems (https://en.wikipedia.org/wiki/Neutron_star#Binary_neutron_star_systems).

Because of the violence of the births of neutron stars in supernova explosions, however, the survival of a binary that contains a neutron star is a complex process that we will not consider further. Suffice it to say that observations demonstrate that binary systems that contain neutron stars actually do survive.

The first neutron-star-containing binary was discovered by Russell Hulse and Joseph H. Taylor at the Arecibo Observatory in 1975. With a period of about 59 milliseconds, PSR 1913+16 is in a very close orbit (the orbital period is less than 8 hours) around an unseen companion of comparable mass (Hulse & Taylor 1975). Einstein's general relativistic theory of gravitation predicts that such massive bodies in such a short-period orbit radiate gravitational waves, and this loss of energy from the orbital motions causes the orbit to decay. Realizing this, Taylor and his collaborators monitored the arrival times of pulses from this object. After a decade and a half, they found the data to be sufficient both to determine the orbital elements and the masses of the pulsar and its unseen companion star and to provide a stringent test of the general relativistic predictions of gravitational radiation (Taylor & Weisberg 1989). They found the pulsar mass to be 1.442 ± 0.003 and the companion mass to be 1.386 ± 0.001 in units of the mass of the Sun, consistent with both objects being neutron stars. They also found that the binary orbit is decaying at a rate equal to 1.01 ± 0.01 times the general relativistic prediction (see Figure 18.5; Wesiberg & Taylor 2005). Hulse and Taylor were awarded the Nobel Prize in 1993 for this discovery.

Although the binary pulsar demonstrated the *existence* of the gravitational radiation predicted by Einstein's theory of general relativity, the actual *detection* of gravitational waves occurred only in 2014, when the Laser Interferometer Gravitational-wave Observatory (LIGO; see Appendix F) reported the first detection of gravitational radiation emitted during the merger of two black holes (see Chapter 19). LIGO is also beginning to enable more accurate determinations of the masses of neutron stars. In 2017, it detected a merger of two neutron stars, which enabled their masses to be determined accurately (Pang *et al.* 2021). And the launch of the Neutron Star Interior Explorer (NICER) on 2017 June 3 initiated an opportunity to use precision X-ray measurements to determine neutron-star radii. An analysis of the gravitational waves emitted during the neutron-star merger GW170817 yielded $R_1 = 10.8^{+2.0}_{-1.7}$ km for the more-massive neutron star and $R_2 = 10.7^{+2.1}_{-1.5}$ km for the less-massive one (LIGO Science Collaboration & VIRGO Collaboration 2018). And combined analyses of the neutron-star mergers GW170817 and GW190425 yielded $R = 11.75^{+0.86}_{-0.81}$ km for a 1.4 M_{Sun} neutron star (Dietrich *et al.* 2020). The most massive neutron star known at the time of this writing is PSR J0740+6620, for which the measured mass is $2.08^{+0.07}_{-0.07}$ M_{Sun} (Fonseca *et al.* 2021) and the radius is $13.7^{+2.6}_{-1.5}$ km (Miller *et al.* 2021).

In 1988, another type of pulsar-containing binary was discovered: PSR B1957+20 is an eclipsing-binary millisecond pulsar, with an unseen, very low-mass companion (roughly 22 times the mass of the giant planet Jupiter; Fruchter *et al.* 1988). Measurements showed that the outflow of energy from the pulsar is slowly ablating the companion. In analogy with black widow spiders that consume their own mates, this object was dubbed the Black Widow Pulsar. As of 2014, some 18 black widow

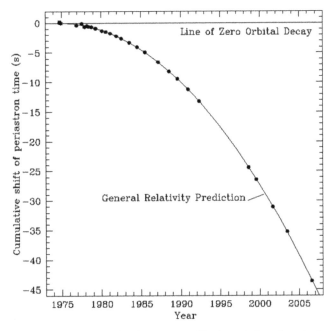

Figure 18.5. Cumulative shift in the time of periastron caused by the emission of gravitational radiation from the orbital motions of the binary pulsar PSR 1913+16 and its unseen companion. Credit: Weisberg, J. M., and Taylor, J. H. (2005); reproduced with permission from the Astronomical Society of the Pacific.

pulsars had been discovered, together with an additional nine "Redback Pulsars," which also are consuming their higher-mass (star-sized) companions.

In 2003, Italian radio astronomer Marta Burgay and her colleagues reported the discovery of the remarkable binary system PSR J0737−3039A/B that consists of two active pulsars (Burgay *et al.* 2003). The primary neutron star in this system is a 22-millisecond pulsar, and the orbital period is only 2.4 hours. The discoverers estimate that the emission of gravitational radiation will cause this system to merge in only about 85 million years.

Another type of neuron-star binary was also discovered in 1975, when Jonathan Grindlay and his colleagues reported the discovery of a burst of X-rays observed with the detectors on board the Astronomical Netherlands Satellite (ANS; Grindlay *et al.* 1976); see Figure 18.6. The two bursts they reported occurred within about 8 hours of each other. Each rose to a maximum within about half a second and then decayed on a timescale of about 8 seconds. The source of the bursts, an object identified as 3U 1820−30, was located close to the center of the globular star cluster NGC 6624. Two years later, Jean Swank and her collaborators at the NASA Goddard Space Flight Center obtained time-resolved X-ray spectra of a burst lasting several minutes from another source (Swank *et al.* 1977). They found the spectra to be fitted best by a blackbody function with temperatures ranging between 0.87 keV (about 10^7 K) to 2.3 keV (about 2.7×10^7 K). The stellar radius they determined from the observations was in the range from about 1 to 10 km, consistent with the X-ray burst source being a neutron star.

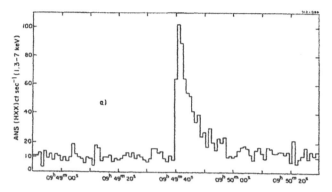

Figure 18.6. The first X-ray burst from a previously identified X-ray source (in the globular cluster NGC 6624) was discovered with the ANS X-ray satellite on 1975 September 28. This luminous burst showed the rapid rise (2 seconds) and exponential decay (10 seconds) typical of the Type I X-ray bursts that were subsequently discovered from many other X-ray sources. Credit: Figure 2(a) from Grindlay, J., *et al.* (1976), reproduced with permission from J. E. Grindlay.

Assuming that these and other similar X-ray bursts might be produced as a result of accretion onto neutron stars, Paul Joss constructed the first detailed models of this type (Joss 1978). If matter of normal composition flows from a companion star onto a neutron star in a sufficiently close binary orbit, then during accretion onto the neutron-star surface, the in-falling matter is heated to temperatures high enough for H-burning. The thermonuclear burning of the accreting matter effectively produces a growing layer of He on the surface of the neutron star. When that layer ultimately becomes thick enough for conditions at its base to ignite He burning via the triple-alpha reactions, Joss's calculations showed that a thermonuclear runaway occurs, producing X-ray bursts that closely resemble those observed.

The X-ray burst sources, which consist of a low-mass ($\leqslant 1 M_{Sun}$) star in orbit with a neutron star, are now known to be a subset of low-mass X-ray binaries (LMXBs). One hundred eighty seven LMXBs were known as of 2007, of which 84 were X-ray bursts sources (Liu *et al.* 2007). These X-ray bursters exhibit both Type I X-ray bursts, which have been ascribed to thermonuclear outbursts involving He burning, as described above, and Type II bursts, which are thought to be accretion events. In 2000, observations of 4U 1735−44 with the Wide Field Camera aboard the BeppoSAX spacecraft—one of a fleet of X-ray satellites that have been deployed to study these and other astronomical sources by various nations since the latter part of the 20th century— revealed a Type I event that decayed slowly over 86 minutes. Dubbed "superbursts," such events have since been observed from six accreting neutron stars, and they have been attributed to unstable thermonuclear burning of the carbon produced during X-ray bursts driven He burning (Cumming & Bildsten 2001; Strohmayer & Brown 2002).

References

Alpar, M. A., & Baykal, A. 2006, MNRAS, 372, 489

Baade, W., & Zwicky, F. 1934, PNAS, 20, 259; see also
   Zwicky, F. 1938, ApJ, 88, 522

Backer, D. C., Kulkarni, S. R., Heiles, C., Davis, M. M., & Goss, W. M. 1982, Natur, 300, 615

Bardeen, J., Cooper, L. N., & Schrieffer, J. R. 1957, PhRv, 108, 1175

Baym, G., Pethick, C., & Pines, D. 1969a, Natur, 224, 673

Baym, G., Pethick, C., Pines, D., & Ruderman, M. 1969b, Natur, 224, 872

Berkowitz, R. 2021, PhT, 74, 15

Burgay, M., et al. 2003, Natur, 426, 531

Chadwick, J. 1932, Natur, 129, 312

Clary, D. 2020, Sci., 370, 1404

Cline, T., et al. 1982, ApJ, 255, L45

Cocke, W. J., Disney, M., & Taylor, D. T. 1969, Natur, 221, 525

Cumming, A., & Bildsten, L. 2001, ApJ, 559, L127

Davies, J. G., Hunt, G. C., & Smith, F. G. 1969, Natur, 221, 27; see also
    Cole, T. W. 1969, Natur, 221, 29
    Radhakrishnan, V., Cooke, D. J., Komesaroff, M. M., & Morris, D. 1969, Natur, 221, 443
    Richards, D. W., & Comella, J. M. 1969, Natur, 222, 551

Dietrich, T., et al. 2020, Sci., 370, 1450

Duncan, R. C., & Thompson, C. 1992, ApJ, 392, L9

Espinoza, C. M., Lyne, A. G., Stappers, B. W., & Kramer, M. 2011, MNRAS, 414, 1679

Fonseca, E., et al. 2021, ApJL, 915, L12

Fruchter, A. S., Stinebring, D. R., & Taylor, J. H. 1988, Natur, 333, 6170

Gold, T. 1968, Natur, 218, 731

Goldreich, P., & Julian, W. H. 1969, ApJ, 157, 869

Grindlay, J., Gursky, H., Schnopper, H., Parsignault, D. R., Heise, J., Brinkman, A. C., & Schrijver, J. 1976, ApJ, 205, L127

Haskell, B., & Sedrakian, A. 2018, ASSL, 457, 401

Hewish, A., Bell, S. J., Pilkington, J. D. H., Scott, P. E., & Collins, R. A. 1968, Natur, 217, 709

https://en.wikipedia.org/wiki/Arkady_Migdal#cite_ref_eduspb_3-0

https://en.wikipedia.org/wiki/Magnetar

https://en.wikipedia.org/wiki/Neutron_star#Binary_neutron_star_systems

Hulse, R., & Taylor, J. H. 1975, ApJ, 195, L51

Joss, P. C. 1978, ApJ, 225, L123

LIGO Science Collaboration and VIRGO Collaboration 2018, PhRvL, 121, 161101

Liu, Q. Z., et al. 2007, A&A, 469, 807

Lorimer, D. R., et al. 2007, Sci., 318, 777

Manchester, R. N. 2018, Pulsar Astrophysics—The Next 50 Years, Proc. IAU Symp. No. 337
    Weltevrede, P., Perera, B. B. P., Preston, L. L., & Sanidas, S. (eds.) (Cambridge: Cambridge Univ. Press) 197

Manchester, R. N., & Taylor, J. H. 1977, Pulsars (San Francisco, CA: Freeman)

Miller, M. C., et al. 2021, ApJL, 915, L28

Nather, R. E., Warner, B., & Macfarlane, M. 1969, Natur, 221, 527

Oppenheimer, J. R., & Volkoff, G. M. 1939, PhRv, 55, 374

Ostriker, J. P., & Gunn, J. E. 1969, ApJ, 157, 1395

Özel, F., & Freire, P. 2016, ARA&A, 54, 401

Pang, P. T. H., et al. 2021, ApJ, 922, 14

Radhakrishnan, V., & Manchester, R. N. 1969, Natur, 222, 228

Reichley, P. E., & Downs, G. S. 1969, Natur, 222, 229

Riley, T. E., et al. 2021, ApJL, 918, L27

Shapiro, S. L., & Teukolsky, S. A. 1983, Black Holes, White Dwarfs, and Neutron Stars (New York: Wiley), 51

Staelin, D. H., & Reifenstein, E. C. III 1968, Sci., 162, 1481; see also
Reifenstein, E. C. III, Staelin, D. H., & Brundage, W. D. 1969, PhRvL, 22, 311

Strohmayer, T. E., & Brown, E. F. 2002, ApJ, 566, 1045

Swank, J. H., Becker, R. H., Boldt, E. A., Holt, S. S., Pravdo, S. H., & Serlemitsos, P. J. 1977, ApJ, 212, L73

Tan, C. M., et al. 2018, ApJ, 866, 54

Taylor, J. H., & Weisberg, J. M. 1989, ApJ, 345, 434

Tolman, R. C. 1939, Phys. Rev., 55, 364

Wesiberg, J. M., & Taylor, J. H. 2005, in Binary Radio Pulsars ed. Rasio, F. A., & Stairs, I. H. ASP Conf. Ser. 328, 25

Chapter 19

Stellar-Mass Black Holes and Gravitational Waves

I was fortunate to be present at the National Science Foundation (NSF) when the LIGO[1] team announced to NSF and the world the first detection of gravitational waves. The presentation was made in the large room where the National Science Board—the presidentially appointed board that oversees the NSF—holds their meetings. Although no one knew exactly what the LIGO representatives were going to say, there was great anticipation that something important was in the works, so the large room was crowded, with standing room only, and an overflow audience was seated in another room with a live television feed.

At the appointed time, NSF Director France Córdova stepped to the microphone and introduced Rainer Weiss from MIT to make the presentation. Also seated in the room at the Board table was Kip S. Thorne, the Caltech theorist who had played major roles both in making the case for funding the LIGO project and in helping to develop the theoretical tools necessary to analyze the results.

Weiss began by explaining the nature of gravitational waves and the structure of the two LIGO detectors, one L-shaped laser interferometer located in Hanford, Washington, and the other in Livingston, Louisiana, some 3000 miles away (Figure 19.1). The LIGO detectors were intentionally constructed in phases, with major efforts required at each stage to reduce the system noise sufficiently to enable the ultimate detection of the extraordinarily faint gravitational-wave signals. The Advanced LIGO detectors finally achieved this level of sensitivity, and they had now made the first-ever direct detection of gravitational waves from a distant astronomical source.

On 2015 September 14, Weiss continued, the LIGO site in Livingston, Louisiana, detected an anomalous signal.[2] Seven milliseconds later, the second LIGO station, in

---

[1] Laser Interferometer Gravitational-wave Observatory; see Appendix F
[2] For additional details, see https://www.ligo.org/science/Publication-GW150914/index.php.

Figure 19.1. Illustration of the LIGO detectors in Hanford, Washington, and Livingston, Louisiana. Source: https://www.ligo.caltech.eedu/assets/what_slide-ce3596915df0767051e5d7d29c27958a.jpg.

Figure 19.2. Rainer Weiss (photo by Bruce Vickman), Barry C. Barish (photo by R. Hahn), and Kip S. Thorne (photo by Keenan Pepper). Source: physicstoday.scitation.or/do/10/1063/pt.6.1.20171003a/fall/.

Hanford, Washington, saw a nearly identical signal. After months of work to verify that the detections were not a manifestation of some sort of noise in the exquisitely sensitive detectors, the LIGO consortium announced on 2016 February 12 that for the first time a gravitational wave produced by an event far out in the universe—dubbed GW 150914—had indeed been directly detected. The signal was independently confirmed by the similar Virgo gravitational-wave detector in Italy, leaving no doubt about the reality of the signal. The discovery was a resounding success for the thousands of scientists and engineers who had been laboring for decades to devise an apparatus capable of detecting the extraordinarily weak ripples in the fabric of spacetime that constitute gravitational waves. Led by Barry C. Barish and Kip S. Thorne of Caltech and Rainer Weiss of MIT (see Figure 19.2), the LIGO team had persevered against seemingly insurmountable odds.

The signal that LIGO detected (Figure 19.3) proved to be a "chirp" of gravitational radiation emitted during the final inspiral of two black holes in a very close binary orbit and their coalescence into a still-larger black hole. Preliminary analyses of the data, including matching the observed signal to templates generated by

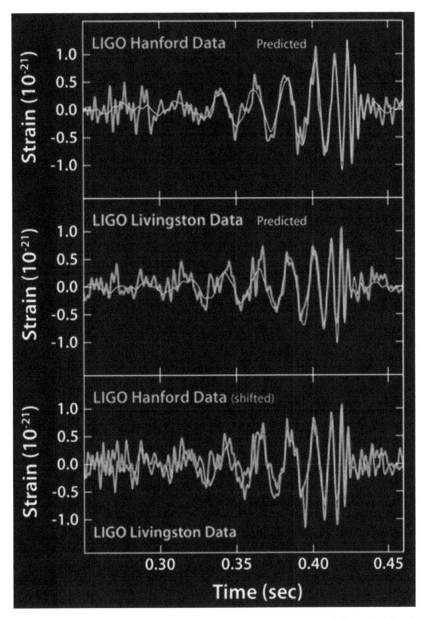

Figure 19.3. The waveform detected for the gravitational-wave signals detected from GW150914. Credit: Caltech/MIT/LIGO Lab.

detailed numerical calculations, indicated that the original binary pair of black holes had masses of about 29 M_{Sun} and 36 M_{Sun}. The more-massive black hole produced by their coalescence was estimated to have a mass of about 62 M_{Sun}, indicating that energy equivalent to about 3 M_{Sun}—some 5×10^{54} ergs!—had been radiated away during the final death spiral of the original pair. This amount of energy is as much as would be radiated by 3000 suns over the entire age of the universe or is equivalent to the amount released in a gigantic supernova explosion of a massive star. Astronomers who were working in wave bands across the electromagnetic spectrum immediately hurried to their telescopes to search for other evidence of this event. Unfortunately, the location of the event in the sky could not be constrained adequately by the limited data available, although it appears to have occurred at a distance of about 1.3 billion lightyears.

Not only did the detection of GW150914 demonstrate the first-ever detection of gravitational radiation, but also it confirmed the existence of black holes. These objects are so dense and their gravitational fields so strong that not even light can escape. That is, the escape velocity from the "surface" of a black hole (more properly called the "event horizon" or the "Schwarzschild radius"[3]) exceeds the speed of light. This surface has a radius of 2.93 km for an object with the mass of the Sun; that is, an object with the mass of the entire Sun packed into a sphere less than 3 km in radius would be a black hole. Such an object would have a density of 1.88×10^{16} g cm^{-3}, or about one hundred times greater than that of an atomic nucleus. The radius of a black hole scales directly with its mass.

The first general-relativistic calculations of the formation of a black hole—though it was not called that at the time—were done by J. Robert Oppenheimer (1904–1967) and Hartland S. Snyder (1913–1962) (Oppenheimer & Snyder 1939). Their analytical work predated the advent of electronic digital computers, but they were nevertheless able to show that, at the end of its thermonuclear lifetime, a star with a mass that is greater than the limiting mass of a neutron star must undergo continuing gravitational collapse. Stellar-mass black holes thus form in core-collapse supernova explosions when the mass of the collapsing core exceeds the TOV limit for the mass of a stable neutron star. For a hypothetical observer co-moving with the collapsing core, the collapse would appear to take about a day. However, as seen by a distant observer, the radius of the star appears to approach the Schwarzschild radius asymptotically as time increases to infinity, while the radiation it emits becomes increasingly red and increasingly faint.

Physicist John Archibald Wheeler (1911–2008) is generally credited with coining the term "black hole" to describe such objects, and the term came into general use after about 1970. The first suspected black hole to be discovered was Cygnus X-1. It was detected as an X-ray source in 1964 during Aerobee rocket X-ray experiments launched by the Naval Research Laboratories (Bowyer et al. 1965). It was subsequently found to be a binary star system, and various estimates of the mass of the invisible X-ray source exceeded the maximum mass of a neutron star, strongly suggesting that it was likely to be a black hole. A recent determination yielded a

---

[3] In 1916, not long before he died from an illness contracted while serving on the Russian front in Kaiser Wilhelm's army during World War I, the outstanding German physicist, astronomer, and mathematician Karl Schwarzschild (1873–1916) obtained the first exact solution to Einstein's general relativistic equations for the gravitational field outside a non-rotating, spherical mass. The solution becomes singular at the Schwarzschild radius $r_s = 2GM/c^2$, where G is the universal gravitational constant, M is the mass of the gravitating body, and c is the speed of light.

mass about 15 times larger than the mass of the Sun, so far above the limiting mass of a stable degenerate star that it effectively eliminates any alternative explanation (Orosz *et al.* 2011).

But, wait a moment! If radiation cannot escape from a black hole, how can it be emitting X-rays? The answer to this question was provided by Soviet scientists Nikolai Shakura and Rashid Sunyaev in 1973 (Shakura & Sunyaev 1973). They pointed out that in a binary star system, matter that is transferred from the normal star to the black hole necessarily carries with it considerable angular momentum. Consequently, it cannot fall directly onto the black hole but instead forms a disk around it. Various physical processes (collectively termed "viscosity") in the disk dissipate the orbital kinetic energy, resulting in the outward transport of angular momentum and the inward flow of matter. The matter is compressed and heated as it flows inward toward the hole, and this heat energy is radiated away as the observed X-ray emission.

Gravitational-wave observations have continued since the discovery of GW150914, and by 2020 October 28, the LIGO/Virgo team had detected 50 gravitational-wave events, which they listed in the second gravitational-wave transient catalog (GWTC-2; Isi & Talbot 2020). These events include GW 170608, the event with the smallest black hole progenitor masses found to date, 10.9 M_{Sun} and 7.6 M_{Sun}; GW 170817, the first detected merger of two neutrons stars; GW 190521, the event with the largest black hole progenitor masses, 85 M_{Sun} and 66 M_{Sun}; and GW 200105, the first detected merger of a neutron star and a black hole (http://en.wikipedia.org/wiki/List_of_gravitational_wave_observations).

Figure 19.4 shows the initial and final (merged) masses for each detection up to 2020, together with the masses of the black holes and neutron stars determined from electromagnetic detections. The greatly increased numbers of detections at the time of this writing also are beginning to enable scientists to determine better the properties of both black holes and neutron stars. For example, they have been able to measure properties such as the spin of a black hole and have found that some spins are misaligned with their orbital angular momenta, raising questions about how these binaries formed. They have also been able to confirm that the remnant black holes behave exactly as expected from Einstein's general theory of relativity. In addition, astronomers had expected to find a mass gap between about 45 M_{Sun} and 135 M_{Sun} due to processes expected to blow apart stars in this mass range before they can collapse into black holes, but the new catalog actually includes mergers within that mass range, presenting theorists with more new challenges (Cho 2020).

Analyses of the gravitational-wave data from LIGO, Virgo, and other existing or planned gravitational-wave detectors depend strongly on theoretical calculations of the emission of gravitational waves. However, the equations of Einstein's general relativistic theory of gravitation are notoriously difficult to solve except for the very simplest geometries. A pair of black holes in a binary orbit—which is known to be a copious source of gravitational radiation—are not among those cases. Consequently, as soon as electronic digital computers began to become widely available in the early 1960s, efforts began to develop numerical methods to solve the equations of general relativity (Arnowitt *et al* 1962; Hahn & Lindquist 1964).

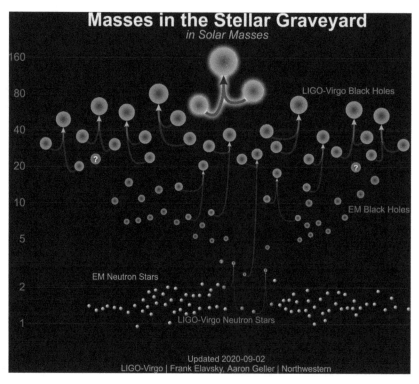

Figure 19.4. Masses in the stellar graveyard. The blue circles are black holes detected by the gravitational radiation emitted during the final inspiral, with the points representing the masses of the two initial black holes connected by yellow arrows to the final, merged black hole. Purple circles represent the masses of black holes detected by electromagnetic radiation (e.g. the X-ray emission from Cygnus X-1); yellow circles represent electromagnetically detected neutron stars, and orange circles represent neutron stars detected by gravitational-wave emission. From https://www.ligo.caltech.edu/news/ligo20201028. Credit: LIGO-Virgo/Northwestern U/Frank Elavsky and Aaron Geller.

By 1975, Larry Smarr and Kenneth Eppley had independently developed numerical codes to follow the evolution of two black holes (Smarr 1975; Eppley 1975). The first realistic calculations of the collapse of a rotating polytropic model, including the emission of gravitational radiation, followed a decade later (Stark & Piran 1985).

The start of LIGO construction in 1994 spurred dramatically increased work on numerical calculations of gravitational radiation from the inspiral and collision of two black holes, but the computing power then available still hampered calculations of three-dimensional geometries. New calculations of the head-on collision of two black holes—the problem studied two decades earlier by Smarr and Eppley—found reasonably good agreement between different approaches to the problem (Anninos *et al.* 1995). Progress thereafter was rapid. In 2001, a team of young scientists at the Max-Planck-Institut für Gravitationsphysik in Germany combined a full numerical solution of the Einstein equations in the strong-field, strongly nonlinear region of the collision with a perturbation approach to compute the gravitational radiation emitted by the resulting highly deformed, rotating black hole to compute the waveforms of the gravitational waves emitted. Calling themselves the "Lazarus Team," the young scientists—John Baker, Bernd Brügmann, Manuela Campanelli, Carlos O. Lousto, and Ryoji Takahashi—succeeded in computing the "plunge waveforms" emitted by an inspiralling pair of black holes (Baker *et al.* 2001). In the early years of the 21st century, they introduced a simple way to fit the plunge waveforms to facilitate the analysis of data from gravitational-wave detectors (Baker *et al.* 2002, 2006). Other investigators were also hard at work on this problem during the same period (Pretorius 2005; Faber *et al.* 2006). The end result of all this effort was that by the time LIGO was able to detect its first gravitational-wave signals, the needed theoretical tools were available to carry out the analysis.

Black holes with masses comparable to those of the stars are not the only ones in the universe. Supermassive black holes (SMBHs)—with masses millions to billions of times larger than the mass of the Sun—are thought to exist in the centers of most, and perhaps all, galaxies (see Chapter 22). At the opposite extreme, black holes that are sufficiently small can actually evaporate over the age of the universe. In 1974, Stephen Hawking (1942–2018) showed that quantum effects—which are not included in Einstein's general relativistic theory of gravitation—actually allow black holes to emit radiation (Hawking 1974, 1988), which has since been named "Hawking radiation." The resulting energy drain causes the black hole to evaporate gradually, but for stellar-mass black holes the lifetime is enormously longer than the 14 billion year age of the universe. However, if very small black holes—with masses less than about 10^{15} g, about the size of a small asteroid—were created in the Big Bang, they would have evaporated by now due to the emission of Hawking radiation.

References

Anninos, P., et al. 1995, PhRvD, 52, 2059

Arnowitt, R., Deseer, S., & Misner, C. W. 1962, Gravitation: An Introduction to Current Research, Witten, L. (ed.) (New York: Wiley) 227

Baker, J., et al. 2001, PhRvL, 87, 121103

Baker, J., et al. 2002, PhRvD, 65, 124012

Baker, J., et al. 2006, PhRvL, 96, 111102

Bowyer, S., Byram, E. T., Chubb, T. A., & Friedman, H. 1965, A&A, 28, 791; see also Bowyer, S., Byram, E. T., Chubb, T. A., & Friedman, H. 1965, Sci., 147, 394

Cho, A. 2020, Sci., 370, 648

Eppley, K. R. 1975, Ph.D. thesis (Princeton Univ.)

Faber, J., et al. 2006, Albert Einstein Century Int. Conf., AIP Conf. Proc. 861,; College Park, MD: AIP) 622

Hahn, S. G., & Lindquist, R. L. 1964, AnPh, 29, 304

Hawking, S. 1974, Natur, 248, 30

Hawking, S. 1988, A Brief History of Time (New York: Bantam Books) 104

http://en.wikipedia.org/wiki/List_of_gravitational_wave_observations

Isi, M., & Talbot, C. M. 2020, https://www.ligo.caltech.edu/news/ligo20201028

Oppenheimer, J. R., & Snyder, H. 1939, PhRv, 56, 455

Orosz, J. A., et al. 2011, ApJ, 742, 84

Pretorius, F. 2005, PhRvL, 95, 121101
Shakura, N. I., & Sunyaev, R. A. 1973, A&A, 24, 337
Smarr, L. L. 1975, Ph.D. thesis (Univ. Texas)
Stark, R. F., & Piran, T. 1985, PhRvL, 55, 891

Part V

Stars with Special Characteristics

The final section of this book begins by relating how scientists have come to understand what causes spontaneous pulsations of different kinds that occur in several different types of stars. Then we discuss the causes of explosive outbursts in some binary star systems. The book concludes with a discussion of our efforts to understand the very first stars to be formed in the universe following the Big Bang. They were very different from the stars around us in the Galaxy today, and efforts to learn more about them are an active area of investigation at the time of this writing.

Chapter 20

Pulsating Stars

Although there had been occasional, sporadic reports of stellar variability before 1784, it seems reasonable to date the study of pulsating stars from that time. In that year, John Goodricke (1764–1786) discovered that the apparent magnitude of the fourth-brightest star in the constellation Cepheus (the King) varies regularly (Pannekoek 1961a; Cox 1980a). Named "δ Cephei"—using the fourth Greek letter to indicate the fourth-brightest star in the constellation after α Cephei, the brightest star—it has since been recognized as the prototype for an entire class of pulsating stars, collectively called "Cepheids," no matter where in the sky they are found. By the end of the 18th century, diligent observations by a number of astronomers had produced a list of 16 variable stars of several different kinds (Cox 1980a). In contrast, by 1969, more than 20,000 variable stars were listed in a catalog for the Milky Way Galaxy (Kukarkin *et al.* 1969).

Beginning in 1843 and continuing over the next 25 years, F. W. A. Argelande (1799–1875) obtained periods and light curves for newly discovered variables (Pannekoek 1961b). By 1892, careful, continuing observations also had demonstrated that the variations in both the luminosity and radial velocity[1] of δ Cephei are periodic (Pannekoek 1961a). Figure 20.1 shows the light curve of δ Cephei plotted over two cycles of pulsation. The observations of the stellar magnitude are folded over the pulsation cycle.

In 1908, Henrietta S. Leavitt (1868–1921), working at the Harvard College Observatory, published a catalog of more than 1700 variable stars in the Magellanic Clouds (Leavitt 1908). In that paper, she made the understated comment, "It is worthy of notice that in Table VI the brighter variables have the longer periods." She subsequently refined this statement in collaboration with Edward C. Pickering (1846–1919), providing a clear plot of the Cepheid period–luminosity relation

---

[1] The radial velocity of a star is its velocity along the line of sight as measured from the Doppler shifts of its spectral lines.

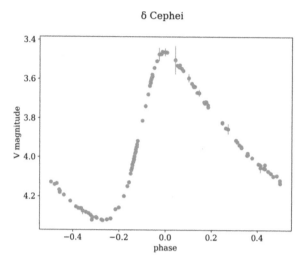

Figure 20.1. Light curve of δ Cephei in the visual (V) Band. Source: commons.wikimedia.org/wiki/File: Delta_cephei_v_engle_2014.png; created by Warrickball 2021 March 16; available under license CC-BY-SA-4.0.

(Leavitt & Pickering 1912). In the late 1920s, this in turn provided part of the foundation for Edwin Hubble's (1889–1953) discovery of the expansion of the universe.

Astronomers in the late 19th and early 20th centuries originally thought that δ Cephei was a binary system because orbital motion is periodic and because it was difficult to imagine that an individual star might actually exhibit time-varying properties. This situation changed completely in 1914, however, when Harlow Shapley (1885–1972) strongly defended the idea that the periodic variations observed in δ Cephei actually are pulsations of the star itself (Pannekoek 1961c; Shapley 1914). Within four years, Arthur Stanley Eddington (1882–1944) produced a theory to account for the pulsations (Eddington 1918). Using the well-established equations of hydrodynamics, he considered purely radial ("breathing mode") adiabatic[2] pulsations of a polytropic stellar model. He pointed out that "a wave that fits the star... determines the natural period of oscillation. For the fundamental oscillation, the first node (or place of constant pressure and density) will fall at the boundary of the star." This is in fact the same condition that determines the fundamental mode of an organ pipe. Eddington's conclusion was that the adiabatic theory of stellar pulsations, as applied to one of Emden's polytropic models, was in fact consistent with the observations of δ Cephei.

Eddington also seems to have been the first to think about possible mechanisms capable of exciting stellar pulsations (Eddington 1926a). He first considered the possibility that the driving force might be associated with the release of subatomic energy (by processes then entirely unknown) near the center of the star. He dismissed this idea for the Cepheids, however, because he considered it to face "fundamental

---

[2] In an adiabatic change, a parcel of matter neither gains nor loses energy.

difficulties." Today, this type of subatomic-driving process is termed the "ε mechanism," with the symbol ε denoting nuclear energy generation. Eddington was right that it does not apply to Cepheids, but it can act as a pulsation-driving mechanism in some other types of stars.

Next, Eddington considered a process that decreases heat leakage during the compression phase of an oscillation and increases it during the expansion phase (Eddington 1926b). He pointed out that for this process to work, "the opacity must increase with compression." He considered this process also to have some difficulties but thought they were "natural to an early stage in the development of a complex theory." Today, this process is called the "κ mechanism," with the symbol κ denoting the opacity of stellar matter, and it was ultimately established as the driving mechanism for Cepheid pulsations. However, at the time of Eddington's early work on this subject, stars were erroneously thought to contain very little hydrogen, so he considered only the opacity due to heavy elements, which retain most of their electrons up to very high temperatures.

A decade and a half later, Eddington returned to the still-unresolved issue of the excitation mechanism for Cepheid pulsations (Eddington 1941). By then, astronomers had realized that hydrogen is the most abundant element in stars (see Chapter 4). Recognizing this, Eddington pointed out that the "critical hydrogen [partial ionization] layer is a comparatively new factor in the problem, since it could not be deemed important until the high abundance of hydrogen in the stars was recognized in 1932... I think it has not hitherto been considered in connection with Cepheid pulsation." He added, "It seems an almost inescapable conclusion that the critical hydrogen layer is the key factor on which... the pulsatory instability and the period–luminosity relation depend."

In the early 1950s, Soviet astrophysicist Sergei A. Zhevakin (1916–2001) suggested instead that the second stage in the ionization of helium (i.e., the ionization of He^+ to He^{++}) might be responsible for driving the pulsations of the Cepheid variables (Zhevakin 1953). In 1959 John P. Cox (1926–1984)—whose 1954 Ph.D. thesis at the University of Indiana had focused on stellar pulsation and who was then a junior faculty member at Cornell University—began a project to investigate this possibility. While previous studies of stellar pulsation had assumed for simplicity that the pulsations were adiabatic, Cox carried out one of the first non-adiabatic calculations of stellar pulsation (Cox 1960). Such calculations are essential for determining whether an oscillation is growing or decaying. He did in fact find a "strong destabilizing influence, resulting from second helium ionization... in the envelope models for ... Cepheids, assuming reasonable helium abundances." In short, Zhevakin was correct, and the mechanism responsible for driving the Cepheid pulsations was finally understood.

Also impeding efforts to understand the nature of Cepheid pulsations was the fact that the internal structures of these stars were largely unknown in the middle of the 20th century. The nature of stellar evolution was just beginning to be understood, and —as we have seen in previous chapters—the internal structures of the stars change greatly during the course of their evolution. The Cepheid variables do not lie on the Main Sequence in the H–R diagram, so their internal structures are necessarily different from those of Main Sequence stars. Indeed, as late as 1962, Norman Baker (1931–2005) and Rudolf Kippenhahn wrote (Baker & Kippenhahn 1962), "As is the case in all previous work on the origin of the Cepheid pulsation, the present investigation suffers from an uncertainty which must remain, so long as the inner structure of the δ Cephei stars is unknown. Only when evolutionary calculations on stellar models shall have been carried into the region of the δ Cephei stars will it be possible to understand ... [them] without need of artificial hypotheses."

Within the decade of the 1960s, the increasing availability of electronic digital computers and of automated methods for calculating stellar evolution changed this situation dramatically. For example, in 1966, Icko Iben Jr. found that his evolutionary models of a 5 M_{Sun} star crossed the observational Cepheid instability strip in the H–R diagram during the phase of core He burning (see Figure 20.2; Iben 1966); the mass of a Cepheid variable is typically 3–6 M_{Sun}. Computer programs were also developed to solve the full set of hydrodynamic equations numerically for spherically symmetric models of the envelopes of the Cepheid variables (Cox et al. 1966).

Astrophysicists' understanding of the cause of stellar pulsation continued to advance steadily, and a decade and a half later, Iben was able to summarize the situation as follows: (Iben 1983) "A star within a narrow strip parallel to and slightly to the left of the line [where hydrogen is 50% ionized at the photosphere] will oscillate at large amplitude in either the fundamental or first overtone radial mode. That it will oscillate is known from the observations. That it oscillates in one of two radial modes (in most cases) is known from a comparison between the results of radial pulsation theory and observation. How a star gets into the instability strip is known from stellar evolution theory. What is responsible for the oscillation is known

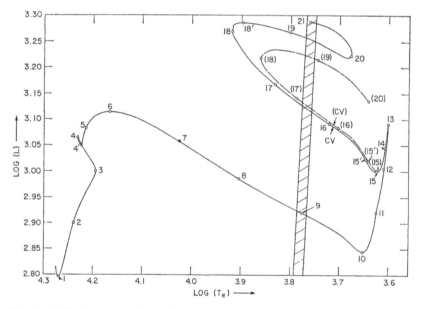

Figure 20.2. Post-Main-Sequence evolution of a 5 M_{Sun} star in the H–R diagram. Source: Iben, I. Jr. 1966, ApJ, **143**, 483; reproduced with permission from I. Iben Jr.

from pulsation theory (driving in the helium and hydrogen ionization zones). The right-hand limit [N.B., the cool edge] to the instability strip is presumably due to the development of a deep convective envelope which damps out the instability mechanism in the ionization zones. The left-hand limit [N.B., the high-temperature edge] is due to the positioning of the driving zones so far below the stellar surface that they cannot 'lift' the layers above them." Although our detailed understanding of the Cepheid variable stars has continued to develop since then, this basic picture remains correct.

20.1 Varieties of Stellar Oscillation Modes

As noted above, the Cepheid variables pulsate in the fundamental mode of radial oscillation; that is, the entire star regularly expands and contracts radially. The outer layers expand and contract most, while the innermost parts of the star vary much less.

Pulsation in the first overtone is also possible. In this case, the outer and inner parts of the star expand and contract in opposition to each other. That is, if the outer layers are expanding, the inner layers are contracting, and vice versa. In a first-overtone pulsation, one shell in the star—called a "node"—does not move. As we shall see below, in another class of pulsating stars, called the RR Lyrae stars, some of the variables pulsate in the fundamental mode, while others pulsate in the first overtone.

Radial pulsations are not the only possible type of stellar oscillation, however. As we saw in Chapter 9, British astrophysicist Thomas G. Cowling (1906–1990) found that there are two general classes of oscillation modes in a non-rotating star, which he termed p-modes and g-modes (Cowling 1941). As for ordinary sound waves, higher p-modes have shorter wavelengths and higher frequencies (shorter periods). In contrast, the fundamental g-mode oscillates at a characteristic frequency called the Brunt–Väsälä frequency, which is essentially the frequency at which a thermally buoyant mass element bobs up and down. Unlike acoustic waves, higher g-mode overtones with shorter wavelengths have lower frequencies (longer periods). Also unlike the p-modes, there are no purely radial g-modes.

The non-radial p- and g-modes have both radial and angular dependences. For a perfectly spherical star, the two are decoupled, and the angular dependence can be expressed in terms of functions called "spherical harmonics" that vary with latitude and longitude on the stellar surface. Like the overtones of the radial functions that can be characterized by the number of radial nodes, the spherical harmonics can be similarly characterized in terms of the number of nodes in latitude (represented by an integer l) and the number of nodes in longitude (represented by an integer m). The first index starts at $l = 0$ (for purely radial modes) and increases in integer steps. For given l, the second index runs from $m = -l$ to $m = +l$ in integer steps. For example, the spherical harmonics with $l = 1$ correspond to dipole modes, in which the "northern" hemisphere of the star oscillates in antiphase with the "southern" hemisphere. Figure 9.1 shows an illustrative example of the surface appearance of a higher-order non-radial mode.

We have already seen that both the Sun and the white dwarf stars undergo multi-mode non-radial pulsations, meaning that they sustain oscillations of many different non-radial modes at the same time.

20.2 The Menagerie of Pulsating Stars

A catalog published in 1919 by Gustav Müller and Ernst Hartwig included 22 β Lyraes, 169 δ Cephei ("Cepheid") variables, and more than 600 Mira variables (Pannekoek 1961b), all recognized as different types of pulsating stars. While theoretical astrophysicists were preoccupied with trying to understand the nature and cause of pulsations in the Cepheid variables, observational astronomers were busily discovering many more types of pulsating stars.

Figure 20.3 shows an illustrative Hertzsprung–Russell diagram with the Main Sequence represented by a short-dashed line and evolutionary tracks for a variety of stellar masses represented by solid lines. Long-dashed lines extending up and to the right from the lower end of the Main Sequence enclose the Cepheid instability strip. The δ Cephei variables occupy the long, cross-hatched oval near the upper end of this strip; we have discussed them in sufficient detail above. The RR Lyrae variables—which we shall see below are low-mass stars with much lower heavy-element abundances than the Sun—lie at lower luminosities, but still within the same instability strip (the small, hatched oval just below the Cepheids). The δ Scuti (δ Sct), rapidly oscillating peculiar A (roAp) stars, and the γ Doradus variables lie near the intersection of the instability strip with the Main Sequence. Mathew Templeton has provided a detailed and very readable summary of these types of pulsating variables (Templeton 2010), and I shall not discuss them further. Below the instability strip along the Main Sequence are multi-mode variables like the Sun. We have already discussed the solar oscillations and helioseismology in Chapter 9, so we shall not consider them further here. The bright Mira variables and irregular (Irr) variables lie along the upper part of the red-giant branch for stars with masses less than about 2 M_{Sun}. Slowly pulsating B (SPB) stars, like 53 Persei, and the β Cephei stars lie at still higher luminosities along the upper Main Sequence.

Finally, along the dotted evolutionary track leading from the high-luminosity phases of stellar evolution down to the white dwarfs we find the variable nuclei of planetary nebulae (or PNNV stars)—also known as GW Virginis stars—and then the DOV, DBV, and DAV white-dwarf variables. Between the Main Sequence and the track leading to the white dwarfs are various hot-subdwarf variables such as the EC 14026 stars. We have discussed the white-dwarf classes of multi-mode pulsators previously in Chapter 15, so we will not consider them further here. The large variety of different types of pulsating stars have recently been discussed extensively by Márcio Catelan and Horace A. Smith, and we refer the interested reader to their work for detailed information (Catelan & Smith 2015). We discuss a few of the other types of variables in a bit more detail below to give the reader a "bird's-eye view" of some of the different types of pulsating stars.

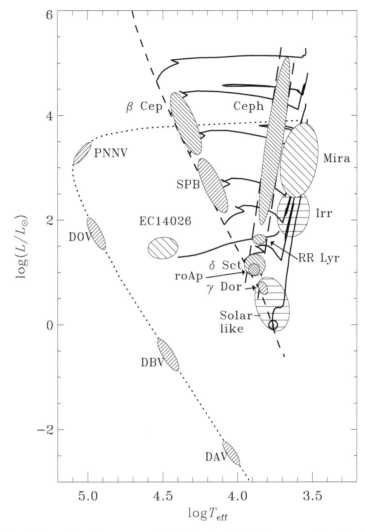

Figure 20.3. Distribution of the different types of pulsating stars in the Hertzsprung–Russell diagram. Source: Christensen-Dalsgaard, J. 1998, in *Proc. First MONS Workshop: Science with a Small Space Telescope*, held in Aarhus, Denmark, June 29–30, 1998, ed. H. Kjeldsen and T. R. Bedding (Aarhus Universiteit), p. 17; articles. adsabs.harvard.edu/full/1998mons.proc...17C; reproduced with permission from J. Christensen-Dalsgaard.

20.2.1 RR Lyrae Variables

The RR Lyrae variables are low-mass stars that lie on the Horizontal Branches of globular star clusters (see Figure 13.5), which are more than 11 billion years old (see Chapter 13), and they have very low abundances of heavy elements. Iben (1983) has described the RR Lyrae stars as follows: "consider those stars in … globular clusters that are burning helium in their cores while on the horizontal branch. Those stars that are at the same time in the instability strip which intersects this branch are known as RR Lyrae stars. They fall into two groups: those with short period[s] and symmetric light curves known as c-type variables and those with longer periods and asymmetric light curves known as ab-type variables. The c-type variables occupy a region in the left hand or hotter portion of the instability strip and the ab-type variables occupy the right-hand portion....[The] variables in the shorter period group are oscillating in the first overtone while those in the long period group are oscillating in the fundamental mode..." The luminosity of the Horizontal Branch—$\log L_{HB} \approx 1.6 \pm 0.1$—and the observed surface temperature of an RR Lyrae star give its radius. Stellar pulsation theory provides a relation between the pulsation period and the stellar mass and radius, which gives the mass of an RR Lyrae star as $M_{RR} \approx 0.6\, M_{Sun}$. This tells us something important about stellar evolution; as Iben points out, "This is substantially less than the mass $M \approx 0.8$ M_{Sun} of a star near the Main Sequence turnoff in a globular cluster. Thus one discovers that, somewhere ... between the core hydrogen burning phase (the Main Sequence) and the core helium burning phase (the horizontal branch), a low-mass star loses a substantial fraction of its original mass." Because "Non-adiabatic pulsation theory provides a composition-dependent estimate of the location of the blue edge of the instability strip..." Iben shows that observations of the RR Lyrae stars constrain their helium abundance to be $Y \approx 0.22 \pm 0.07$. He concludes, "Thus, with the help of the observational and theoretical properties of RR Lyrae stars, we have been able not only to establish the existence of a mass loss process (which theory cannot as yet produce from first principles) but have been able to show that the abundance of helium in the oldest stars is substantially greater than zero..." which is consistent with the calculations of nucleosynthesis in the early universe based on the 3 K temperature of the cosmic microwave background radiation.

20.2.2 Solar-like Variables

In the early 1990s, Swiss astrophysicist Didier Queloz, working under the direction of his mentor Michel Mayor, constructed a new type of spectrograph capable of measuring the radial-velocity variations of stars to unprecedented accuracy. Their goal was to use this instrument to search for the very small, periodic variations in the radial velocity of a star that would be produced by an extrasolar planet ("exoplanet") in orbit around it. In 1995, they succeeded in detecting the first exoplanet, 51 Pegasi b (Mayor & Queloz 1995).[3] In 2019 they were awarded a share in the Nobel Prize for this work.

This discovery sparked an extensive search for other exoplanets; as of this writing more than 4000 are known. However, the search for very low-level radial-velocity and intensity variations also made it possible to search for low-level solar-type variability in other stars, and by the end of the 1990s solar-like oscillations were being found in other Main Sequence and subgiant stars. In 2002, such oscillations were detected in the G7 giant star ξ Hydrae (Frandsen *et al.* 2002). This 3.07 M_{Sun} star has left the Main Sequence and is in the stage of H burning in a shell, and it has now reached the very base of the red-giant branch. It sustains multiple radial p-mode

---

[3] Astronomers have long used capital letters to designate binary companions; for example, the faint white dwarf Sirius B is in a binary orbit with the bright star Sirius (renamed Sirius A after its binary companion was discovered). Astronomers accordingly adopted lower-case letters to designate planetary companions.

oscillations with periods between 2.0 and 5.5 hours. The asteroseismological constraints also enabled the investigators to improve the measured values for stellar parameters such as the effective temperature and luminosity.

Beginning in the first decade of the 21st century, spacecraft observations dramatically increased the numbers of stars with new asteroseismological constraints. *CoRoT* (for Convection, Rotation, and Planetary Transits) was a French spacecraft specifically designed both to search for exoplanets and to perform asteroseismological investigations. Launched by the European Space Agency (ESA) in 2006, the satellite contributed significant discoveries in both areas during its seven-year lifetime. Among other things, it obtained the first detection of non-radial oscillations in red giants, found solar-like oscillations in massive stars, discovered hundreds of frequencies in δ Scuti stars, and observed *g*-mode oscillations in slowly pulsating B stars[4]. The Kepler spacecraft launched by NASA in 2009 also contributed substantially to astronomers' knowledge of the solar-like oscillations of other stars (Chaplin *et al.* 2011). It detected such oscillations in some 500 Main Sequence and subgiant stars before the mission was finally terminated in 2018. As one group of astronomers put it, "Asteroseismology of red giant stars has been one of the major successes of the *CoRoT* and *Kepler* missions" (Aguirre *et al.* 2020). Since 2018 these missions have been superseded by the *Transiting Exoplanet Survey Satellite* (*TESS*). Although its primary mission was to search for exoplanets, it is also capable of obtaining precision photometric data for hot stars. As of 2020, it had provided measurements of the global asteroseismic observables for some 25 stars and improved estimates of their fundamental stellar properties.

20.2.3 Mira Variables

In 1596, the Dutch astronomer David Fabricius (1564–1617) discovered a variable star with a very long period, roughly 11 months. Now known as omicron Ceti (o Ceti = Mira), it is the primary star in a binary system that includes a white dwarf and which is located about 92 parsecs away in the constellation Cetus (the Whale). Mira undergoes regular pulsations with a period of 332 days, and it is the prototype of the class of long-period variables that bear its name. Figure 20.4 shows a portion of the visual light curve of Mira (Huang *et al.* 2018) obtained from decades of observations by members of the American Association of Variable Star Observers (AAVSO). Founded in 1911, the AAVSO has since "coordinated, collected, evaluated, analyzed, published, and archived variable star observations made largely by amateur astronomers and makes the records available to professional astronomers, researchers, and educators." (https://en.wikipedia.org/wiki/American_Association_of_Variable_Star_Observers) Members of the AAVSO are amateurs only in the sense that they are not paid for doing the observing that they love. Indeed, many obtain their observations with great skill using sophisticated equipment.

Mira variables are very red (with $T_{eff} \sim 3000$ K), very large ($R \sim 200$–300 R_{Sun}), and very luminous ($L \sim 3000$–4000 L_{Sun}; Mattei 1997; Iben 1983). They are AGB

---

[4] See the special October 2009 issue of Astronomy & Astrophysics, volume 506.

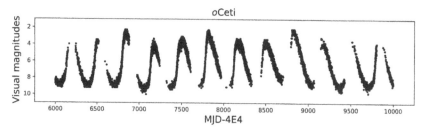

Figure 20.4. A portion of the visual light curve of the long-period variable Mira. Source, Huang, C. D., *et al.* (2018), figure 4; reproduced with permission from C. D. Huang.

stars less massive than about 2 M_{Sun}, and they are currently burning H and He in concentric shells sources around a dense and degenerate carbon–oxygen core. They are currently undergoing rapid mass loss, at rates of about 10^{-6} M_{Sun} yr^{-1}. They pulsate with large amplitudes (more than 2.5 magnitudes in the visible), with periods ranging from about 150 days to 1000 days. The AAVSO astronomers have been monitoring Mira variables for more than 80 years and have compiled an extensive database of light curves for these and other variable stars (Watson *et al.* 2006). Miras are divided into O-rich and C-rich stars. The former have periods ranging from about 100 to about 450 days, with most clustering near 200 days. In contrast, the C-rich Miras are more luminous and have periods ranging from about 250 to about 600 days (Huang *et al.* 2018).

Because of the very low densities of their highly distended envelopes, the very long periods of the Mira variables are not surprising if the pulsations are *p*-mode oscillations (including radial pulsations). The time required for an acoustic wave to traverse a star can be shown to be inversely proportional to the square root of its mean density (Cox 1980b). Thus, a star with about the mass of the Sun but with a radius some 200–300 times larger would be expected to have a pulsation period of the order of a few hundred days. The oscillations are thought to be driven by the κ mechanism operating in the partial-ionization zone of hydrogen (Iben 1983), although turbulent pressure in the deep convection zone may also contribute (Xiong *et al.* 2016).

One or two percent of the Mira variables exhibit dramatic changes in their pulsation periods. For example, the Mira variable T Ursa Minoris (T UMi) had a pulsation period of about 315 days before 1970, but it had declined to only 114 days by 2008 (Fadeyev 2018). Such a change is generally thought to be due to a thermal pulse of the He-burning shell in the AGB star (Smith 2013; see Chapter 13).

20.2.4 β Cephei Variables

In the early 1900s, Dartmouth College astronomer Edwin B. Frost (1866–1935) discovered spectroscopically that the radial velocity of the star β Cephei varies with a period of about 5 hours. He interpreted this to mean that the star was a binary. However, a decade later, Paul Guthnick (1879–1947) at the Berlin Observatory found that β Cephei also exhibits low-level variability—less than one hundredth of a magnitude in the visual-wavelength band—with the same period. The star was thus shown to be a pulsator rather than a binary (Percy 2010). It is in fact the type member of the class of pulsating stars that bears its name. Also sometimes called β Canis Majoris variables, they are extremely hot, with effective temperatures of 20,000–30,000 K, and they are all low-amplitude pulsators with oscillation periods of 3.5–6 hours. They generally are slow rotators, and they are confined to a narrow region in the H–R diagram that coincides closely with the post-Main-Sequence turnaround in the evolutionary tracks of 10–20 M_{Sun} stars, "where the stellar structure changes from core hydrogen burning to overall contraction and then to shell hydrogen burning" (Unno et al. 1979a).

By 1967, 41 β Cephei stars were known (Percy 1967). About half exhibited beats or multiple periodicities. The beats have usually been interpreted as the result of interference between two oscillations with nearly equal periods (Unno et al. 1979b). In 1951, the distinguished Belgian astrophysicist Paul Ledoux (1914–1988) suggested that these observations could be accounted for if the pulsations are non-radial oscillations of the star. Two decades later, Japanese astrophysicist Yoji Osaki calculated line profiles for a slowly rotating star undergoing such non-radial oscillations. Assuming that the spectral lines are split by the rotation of the star, he also estimated equatorial rotation velocities from the beat periods. Obtaining satisfactory agreement with rotation velocities determined from line widths, he concluded that non-radial p-mode oscillations can indeed explain most of the fundamental characteristics of the β Cephei stars (Osaki 1971).

Understanding the mechanism responsible for driving the oscillations in the β Cephei stars perplexed astronomers and astrophysicists for most of the rest of the 20th century, however. Although, as we have seen, the excitation mechanism for the classical δ Cepheid variables had been identified in the 1950s with the He^+ ionization zone, the β Cepheids are far too hot for that. Both H and He are almost completely ionized even at the photospheres of these stars. A glimmer of hope appeared in 1978, when Robert F. Stellingwerf showed that, while He^+ ionization does tend to destabilize a β Cepheid model, it was necessary to increase the opacity artificially in order to excite oscillations (Stellingwerf 1978). Meanwhile, other problems had accumulated as astronomers worked to match theory and observations for the classical δ Cepheid variables, and this led Norman R. Simon to issue a plea for a reexamination of the contributions of heavy elements to stellar opacities (Simon 1982).

For decades, astrophysicists had been using opacities generously provided to the community by Arthur N. Cox (1927–2013) and his colleagues at Los Alamos National Laboratory (LANL). The LANL scientists had originally developed computer programs to compute the radiative opacities of various chemical elements and mixtures at very high temperatures and densities in order to model the development of a nuclear explosion. Because stellar interiors have similarly high temperatures, it was natural for Cox and others to extend them to astrophysical problems as well. Cox described the methods of calculation in the mid-1960s Cox 1964, and he and his colleagues published various tables of opacities as functions of temperature and density for astrophysical use (Cox et al. 1965).

These opacities had subsequently become the workhorses for calculations in stellar astrophysics.

In response to Simon's plea, however, scientists at both LANL and the Lawrence Livermore National Laboratory (LLNL)—which had been established as an independent national laboratory in 1952 to focus on the development of thermo-nuclear weapons—reexamined the opacity calculations. The LLNL scientists found that some small atomic-physics interactions, which had been neglected in the Los Alamos calculations, in fact are not negligible for heavier elements such as iron (Fe). This allows a multitude of radiative transitions, which increase the total opacity of an elemental mixture by as much as a factor of three at temperatures of the order of 300,000 K and densities like those in stellar envelopes (Iglesias *et al.* 1992). The Livermore scientists termed their new results the "OPAL" ("OPacities At Livermore") opacities, and the localized increase in opacity at temperatures near 300,000 K came to be known as the "Fe bump" or the "Z bump," with the symbol Z commonly used by astronomers and astrophysicists to refer collectively to all the elements heavier than H and He. Meanwhile, an international collaboration of astrophysicists working independently of the American groups produced a third set of opacity calculations. Led by the British astrophysicist Michael J. Seaton (1923–2007), the Opacity Project (OP) produced detailed Rosseland mean opacities for mixtures of elements under conditions appropriate to stellar envelopes (Seaton *et al.* 1994). The OP opacities generally agree well with the OPAL opacities, although some differences do exist. Both sets of calculations yield larger Z-bump opacities than the earlier LANL calculations.

It did not take long for theorists around the world to exploit the new opacities and demonstrate that they do indeed result in pulsational instabilities in the O and B stars (Cox *et al.* 1992). The long-standing problem of the mechanism responsible for exciting the pulsations in the β Cephei was thus solved thanks to the interplay between astronomy and physics that led to improvements in our understanding of these fundamental physical quantities.

References

Aguirre, V. S., et al. 2020, ApJL, 889, L34

Baker, N., & Kippenhahn, R. 1962, ZA, 54, 114

Catelan, M., & Smith, H. A. 2015, Pulsating Stars (Weinheim: Wiley-VCH)

Chaplin, W. J., et al. 2011, Sci., 332, 213

Cowling, T. G. 1941, MNRAS, 101, 367

Cox, A. N., Brownlee, R. R., & Eilers, D. D. 1966, ApJ, 144 1024; see also
 Cox, J. P., Cox, A. N., Olsen, K. H., King, D. S., & Eilers, D. D. 1966, ApJ, 144, 1038

Cox, A. N. 1964, JQSRT, 4 737; see also
 Cox, A. N. 1968, Stellar Structure: Stars and Stellar Systems, vol. 8, Aller, L. H., & McLaughlin, D. B. (eds.) (Chicago, IL: Univ. Chicago Press) 195

Cox, A. N., Morgan, S. M., Rogers, F. J., & Iglesias, C. A. 1992, ApJ, 393 272; see also
 Kiriakidis, M., El Eid, M. F., & Glatzel, W. 1992, MNRAS, 255, 1P
 Moskalik, P., & Dziembowski, W. 1992, A&A, 256, L5
 Dziembowski, W., & Pamyatnykh, A. A. 1993, MNRAS, 262, 204

Cox, A. N., Stewart, J. N., & Eilers, D. D. 1965, ApJS, 11 1; see also
　　Cox, A. N., & Stewart, J. N. 1965, ApJS, 11, 22

Cox, J. P. 1960, ApJ, 132, 594

Cox, J. P. 1980a, Theory of Stellar Pulsation (Princeton, NJ: Princeton Univ. Press) 5

Cox, J. P. 1980b, Theory of Stellar Pulsation (Princeton, NJ: Princeton Univ. Press) 10

Eddington, A. S. 1918, MNRAS, 79, 2

Eddington, A. S. 1926a, Internal Constitution of the Stars (Cambridge: Cambridge Univ. Press),
　　200

Eddington, A. S. 1926b, Internal Constitution of the Stars (Cambridge: Cambridge Univ. Press)
　　202

Eddington, A. S. 1941, MNRAS, 101, 182

Fadeyev, Y. A. 2018, AstL, 44, 546

Frandsen, S., et al. 2002, A&A, 394, L5

https://en.wikipedia.org/wiki/American_Association_of_Variable_Star_Observers; see also www.
　　aavso.org.

Huang, C. D., et al. 2018, ApJ, 857, 67

Iben, I. Jr. 1966, ApJ, 143, 483

Iben, I. Jr. 1983, SoPh, 82, 457

Iglesias, C. A., Rogers, F. J., & Wilson, B. G. 1992, ApJ, 397 717; see also
　　Rogers, F. J., & Iglesias, C. A. 1992, ApJS, 79, 507

Kukarkin, B. V., et al. 1969, General Catalogue of Variable Stars, Volume 1: Constellations
　　Andromeda—Grus (Moscow: Astronomical Council of the Academy of Sciences of the
　　USSR: Sternberg State Institute of the Moscow State Univ.)

Leavitt, H. S. 1908, AnHar, 60, 87

Leavitt, H. S., & Pickering, E. C. 1912, HarCi, 173,

Mattei, J. A. 1997, JAVSO, 25, 57

Mayor, M., & Queloz, D. 1995, Natur, 378, 355

Osaki, Y. 1971, PASJ, 23, 485

Pannekoek, 1961a, A History of Astronomy (London: George Allen & Unwin Ltd.); reprinted
　　1989 (New York: Dover) 319

Pannekoek 1961b, A History of Astronomy (London: George Allen & Unwin Ltd.) 435

Pannekoek, 1961c, A History of Astronomy (George Allen & Unwin Ltd.) 439

Percy, J. R. 1967, JRASC, 61, 117

Percy, J. R. 2010, https://www.aavso.org/vsots_betacep

Seaton, M. J., Yan, Y., Mihalas, D., & Pradhan, A. K. 1994, MNRAS, 266, 805

Shapley, H. 1914, ApJ, 40, 448

Simon, N. R. 1982, ApJ, 260, L87

Smith, H. A. 2013, in 40 Years of Variable Stars, arXiv:1310.0533, Kinemuchi, K., Smith, H. A.,
　　De Lee, N., & Kuehn, C. (eds.)

Stellingwerf, R. F. 1978, AJ, 83, 1184

Templeton, M. 2010, https://www.aavso.org/vsots_delsct

Unno, W., Osaki, Y., Ando, H., & Shibahashi, H. 1979a, Nonradial Oscillations of Stars (Tokyo:
　　Univ. Tokyo Press) 23

Unno, W., Osaki, Y., Ando, H., & Shibahashi, H. 1979b, Nonradial Oscillations of Stars (Tokyo:
　　Univ. Tokyo Press) 26

Watson, C. L., Henden, A. A., & Price, A. 2006, 25th Annual Symp. on Telescope Science SASS 25, 47

Xiong, D. R., et al. 2016, MNRAS, 457, 3163

Zhevakin, S. A. 1953, AZh, 30, 161

Chapter 21

Cataclysmic Variables

In contrast to pulsating stars, which exhibit periodic oscillations, cataclysmic variables (CVs) instead undergo outbursts during which the intrinsic luminosity of the star temporarily increases greatly before gradually declining back to a lower level. Recorded observations of CVs extend far back into antiquity. Chinese records of "guest stars" exist back to about 200 BCE, and Japanese and Korean records go back to about 800 CE (Stephenson 1976). However, these "guest stars" included comets as well as objects that we now call supernovae, novae, and dwarf novae. Conversely, except for the supernova of 1006, useful observations of CVs in Europe and the Arab world date only from the Renaissance.

Gradually, several different types of CVs have come to be identified. They range from various kinds of dwarf novae, which display only moderate increases in brightness, to classical novae, to supernovae, and to the much more recently discovered extraordinarily powerful hypernovae. The cause of each type of outburst has proven to be qualitatively different from the others. The purpose of this chapter is not to provide a detailed description of all the different types of CVs but instead to give a very broad-brush overview of some of the main classes. Those who may be interested in more detail are referred to the comprehensive review by Warner (2003).

21.1 Classical Novae

Because the outburst of a nova cannot be predicted—unlike the regular oscillations of a pulsating star—astronomers have no choice but to wait for one to appear and then to concentrate on it using the best available techniques. By observing many different novae over the course of the rapid rise and more gradual decline in their light output, by 1950 astronomers had been able to piece together a general picture of a nova outburst. Dean B. McLaughlin (1901–1965) summarized the main results from the accumulated observations of novae as follows: (McLaughlin 1950) (1) The star that is the seat of the nova eruption ejects a shell of gas that expands outward at an approximately constant speed and then gradually fades away. (2) The remnant star that is left behind is a hot, dense star with a radius of about 0.1 R_{Sun} and an effective temperature of about 50,000 K. (3) The post-nova remnant differs very little from the pre-nova star; in particular, the ejected mass is very much less than the mass of the star. As illustrated in Figure 21.1 (see also McLaughlin 1943), for our present purposes a nova outburst can be considered to consist of a very rapid rise to a very high luminosity, generally in about a day, followed by a gradual decline back to the pre-eruption level on a timescale that may be as long as a month or more.

A major step toward understanding the nature of classical novae resulted from the 1954 discovery by Merle F. Walker that Nova DQ Herculis (1934) is a binary star (Walker 1954). This prompted astronomers to wonder whether all novae might be double stars. Robert P. Kraft (1927–2015; Figure 21.2), then an Assistant Professor at Indiana University (Kraft 2009), showed that the recurrent nova T Coronae Borealis (T CrB), which had been observed to undergo nova explosions in both 1866 and 1946, is a double-lined spectroscopic binary with an orbital period of 227.6 days (Kraft 1958a). One component of the binary is a red giant star of spectral type gM3, while the other is a hot blue, star. According to Kraft, "It is very likely that the gM3 star fills one lobe of the inner zero-velocity surface defined by the restricted three-body problem." Figure 21.3 shows an artist's conception of a binary star nova system like the one Kraft proposed for T CrB. The red giant at the left in this figure has expanded to fill its Roche lobe (the "inner zero-velocity surface" mentioned by Kraft). This is the greatest extent to which the gaseous outer layers of the red giant can expand while still remaining bound to the star by gravity. Because the companion star in the binary system—the bright, blue-white area at the right in this image—also exerts a gravitational pull on the gas, the surface of the distended red giant is pulled into a teardrop shape, with the narrow end pointing toward its companion. As the evolution of the red giant causes it to try to expand further, gas from it expands through the narrow point of the teardrop shape (called the "inner

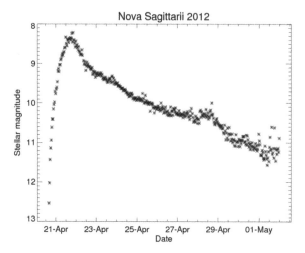

Figure 21.1. Light curve of Nova Sagittarii 2012, an example of a classical nova outburst. Credit: William T. Thompson, courtesy of NASA.

Figure 21.2. Robert P. Kraft in June, 1969. Credit: AIP Emilio Segrè Visual Archives, John Irwin Slide Collection.

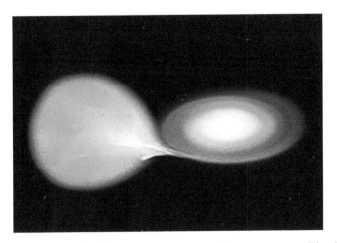

Figure 21.3. Artist's conception of a binary star nova system. Source: commons.wikimedia.org/wiki/File: Making_a_Nova.jpg. Credit: NASA/CXC/M.Weiss.

Lagrangian point"), forming a stream that falls into the Roche lobe centered on the blue-white star to the right. The angular momentum carried by the stream prevents it from falling directly onto the surface of the companion, and instead it circles around it to form a gaseous disk. Dissipation within the disk—termed an "accretion disk" by astrophysicists—enables matter to flow gradually inward and accumulate on the surface of the companion. Kraft's observational evidence supported the concept that material flows from the red star through the inner Lagrangian point to form a "ring or shell-like region" around the small, hot, blue star. He also noted that dwarf novae of the U Gem class—to be discussed below—have a physical structure similar to that of T CrB, and he suggested "that the luminosity of the blue components of these systems is maintained by accretion heating, the material being supplied by the red companions." Indeed, he found that the blue components of T CrB and of the dwarf novae SS Cyg and AE Aqr are all brighter than typical white dwarfs, consistent with the idea that they are white dwarfs that are experiencing accretion-powered heating.

Kraft continued his studies of novae after he had accepted a position at the Mount Wilson and Palomar Observatories, and he and others published an extended series of papers on this subject under the generic title "Binary Stars Among Cataclysmic Variables." His 1964 paper (Kraft 1964) is often cited as providing compelling evidence that all novae likely involve outbursts on white dwarfs in close binary systems. The late Polish theorist Bohdan Paczyński (1940–2007) summarized the situation as of 1965: (Paczyński 1965) "The recent discoveries of many binary systems among old novae ... seem to leave no doubt ... [about the] association between the nova phenomenon and evolution of some types of close binary systems." After reiterating the general picture laid out by Kraft, he went on to say that the "most generally accepted theory of outbursts is due to Schatzman... (Schatzman 1951). According to his theory, an outburst can occur when energy sources are close to the stellar surface..." He concluded that "It is very probable that during the process of mass accretion [onto the largely degenerate blue star] some critical temperature or density will be reached at the base of the hydrogen-rich envelope and a nova outburst will occur."

The observations thus suggested that a nova outburst involves a thermonuclear explosion on a white dwarf in a close binary system. The peak luminosity of a nova may be roughly 10,000 times its pre-eruption value. As the luminosity of the system prior to eruption may be comparable to the solar luminosity, then if the decline back to the pre-outburst level takes about a month, the total energy radiated during that time is of the order of 10^{44} ergs. This is about as much as the Sun radiates in 2000 years! Clearly, a substantial source of energy is required to power such an eruption. In the middle of the 20th century, astronomers were unsure what that energy source was, although a few, like French astrophysicist Evry Schatzman (1920–2010), had begun to consider shock waves powered by thermonuclear explosions (Schatzman 1949[1]).

The obvious next step was to carry out hydrodynamic calculations capable of following the explosive ignition of H burning in the layer of H-rich gas accreted onto the surface of a degenerate white dwarf. This seems to have been done first by Sumner Starrfield, James W. Truran (1940–2022), Warren M. Sparks, and G. Siegfried Kutter in 1972 (Kutter et al. 1972). They did not follow the accretion process itself but instead placed a spherical shell of gas artificially on the surface of the white dwarf. After ensuring that it was initially in both hydrostatic and thermal

---

[1] T. A. Eminadze (1962, SvA, 6, 431) also credits L. E. Gurevich and A. J. Lebedinski as proposing "as early as 1947... [that] the nova results from a thermal explosion due to nuclear reactions taking place in the peripheral layers of a star."

equilibrium, they followed the subsequent evolution with a hydrodynamic computer code. If the shell contained sufficient mass, its deepest layers were hot and dense enough to undergo thermonuclear H burning.

In one model constructed by Starrfield and his colleagues, the white dwarf had an initial luminosity of 0.2 L_{Sun} and a radius of 8200 km, and its effective temperature was 35,000 K. For an envelope mass of about 1.25×10^{-3} M_{Sun}, the temperature and density at the base had become compressed to 2.96×10^{7} K and 1.27×10^{4} g cm^{-3}, respectively, making that layer strongly degenerate. Initially, the H-burning p–p reactions dominated. Because the bottom of the accreted layer was degenerate, initially it did not expand, and the temperature continued to increase. However, in only two days the temperature had surged to 9×10^{7} K, removing the degeneracy and thus allowing the envelope to expand; about half the envelope had become convective; and the CNO-cycle reactions had taken over. In only 95 seconds more, the thermonuclear reaction rate began to drop rapidly because all the available CNO nuclei had been converted into β^{+}-unstable nuclei, and the relatively long β-decay timescales prevented further reactions.

However, Starrfield and his colleagues found that models with normal CNO abundances did not produce outbursts. Accordingly, they artificially enhanced the abundances in the deepest layers to 40% of the mass—to mimic, e.g., matter from the C/O core of the white dwarf being "splashed up" into the surface layers during accretion. In these cases, they found that the rapid rise in the release of thermonuclear energy caused the pressure in the deepest layers of the accreted envelope to spike, driving shock waves both into the core and out toward the surface. When the shock wave reached the surface, it ejected 1.5×10^{-5} M_{Sun} at a speed of 7400 km s^{-1}. Some of the spherically symmetric models studied by these investigators reached absolute visual magnitudes peaking around $M_V \approx -5$ or -6, and the ejected masses carried kinetic energies of 8×10^{44} ergs.

Since these early investigations, various hydrodynamic studies have shown that the amount of matter that can be accreted onto a white dwarf depends on its mass and luminosity, the accretion rate, and the composition of the accreted matter (Starrfield *et al.* 2016). The higher the initial white-dwarf luminosity, the smaller the amount of mass accreted. If the accretion rate exceeds 10^{-7} M_{Sun} yr^{-1}, the accreting material is hot enough to burn steadily as it accumulates on the white dwarf, and there is no thermonuclear outburst. However, when the initial luminosity is less than 10^{-2} L_{Sun} and the accretion rate is less than 10^{-9} M_{Sun} yr^{-1}, a layer of some 10^{-6} to 10^{-4} M_{Sun} of unburned H-rich matter gradually accumulates on the white dwarf.

This picture of a nova explosion, originally developed by Starrfield and his colleagues nearly half a century ago, is generally accepted today. However, some problems still remain. For example, what actually causes the significant enrichment of the CNO abundances that is necessary to produce high levels of thermonuclear energy generation on hydrodynamic timescales? It presumably originates from matter dredged up from the degenerate C/O core of the white dwarf, but how and when does this occur? Various possibilities have been suggested, but as of this writing the issue remains unresolved. Because the accretion flow onto the white-dwarf surface is rotating at almost the Keplerian orbital velocity, another possibility is that the strong shear between the accretion flow and the surface may drive shear mixing (Kippenhahn & Thomas 1978). Still another possibility is that strong convection driven by the thermonuclear runaway may dredge up material from the white-dwarf core (Glasner & Truran 2012). Or mixing may be due to a Kelvin–Helmholtz instability (Casanova *et al.* 2011). In addition, while the bolometric light curves and nucleosynthetic abundances produced by the thermonuclear-runaway model are in reasonable agreement with observations (Starrfield *et al.* 1998), a further problem is that the ejected masses predicted by theoretical calculations are as much as 10 times smaller than the observations show: about 2×10^{-4} M_{Sun} of matter is observed to be ejected from the white dwarf during the outburst (Gehrz *et al.* 1998).

21.2 Dwarf Novae

On 1855 December 15, English astronomer John Russell Hind (1823–1895) happened to observe the beginning of an outburst in an object we now call U Geminorum (Hind 1856). He observed it again on 1856 December 18 and 1856 January 10, as it faded from a very blue, ninth-magnitude star to one of twelfth magnitude. This "very interesting" variable star became the prototype of a class of CVs that are now called dwarf novae. Forty years later, Louisa D. Wells, working at the Harvard College Observatory, discovered another object of this type, now known as SS Cygni (Unknown 1896). Additional members of this class were subsequently discovered, and by 1948 more than 50 had been identified (McLeod 1948). A dwarf nova is mostly a faint object, but occasionally it increases abruptly in luminosity by a factor of perhaps 10 to 100, remains at this higher level for a day or two, and then fades back to its original luminosity over a few days (see Figure 21.4). The outbursts generally recur at intervals of weeks to months. The amount of energy involved in a dwarf nova

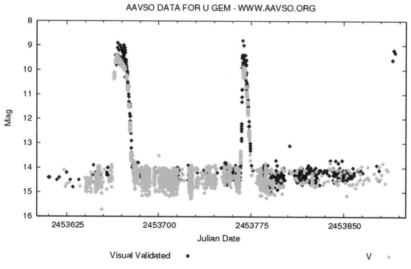

Figure 21.4. Light curve of the dwarf nova U Geminorum during outbursts. Generated from http://www. aavso.org/data/lcg and used with permission. Source: commons.wikimedia.org/wiki/File:U_Gem_aavso.png.

outburst thus is much less than that released in a classical nova outburst, so it is very unlikely to be a thermonuclear event. But what is the energy source?

From observations over the decade 1943 to 1953, Alfred H. Joy (1882–1973) at the Mount Wilson and Palomar Observatories found that the dwarf nova AE Aquarii is a double-lined spectroscopic binary consisting of a red star and a faint, blue star (Joy 1954). The following year, Berkeley astronomers John A. Crawford (1921–2010) and Robert P. Kraft, then a predoctoral fellow, observed this system again at the Lick Observatory and found the red object to be a star of spectral type K5 in an early post-Main Sequence stage of evolution. They proposed a model for the system in which the red star fills its Roche lobe and transfers mass over to its small companion. This model is virtually the same as that advanced by Kraft two years later for Nova T Coronae Borealis. Crawford and Kraft suggested that "accretion of mass by the blue star is related to the [observed] outbursts..." and that the system may be related to the U Geminorum type variable (Crawford & Kraft 1956). The term "dwarf nova" seems first to have been introduced for these types of cataclysmic variables in 1959 by Jesse L. Greenstein (1909–2002) at Caltech and by Kraft, then at the Yerkes and McDonald Observatories (Greenstein & Kraft 1959).

The nature of dwarf-nova outbursts—and even the site of the eruption—remained uncertain for some time. Did the eruption take place on the cool, red star in the binary system (Krzeminski 1965)? Were they small-scale thermonuclear events like miniature novae that occurred on the white dwarf (Saslaw 1968)? Was the outburst due to unstable, episodic mass loss from the red component (Bath 1969, 1972)?

Around this same time, during the late 1960s and early 1970s, astronomers were also beginning to recognize that mass transfer in a close binary system necessarily carries with it angular momentum. Consequently, matter does not fall directly onto the compact companion but instead forms a disk surrounding it. Matter gradually flows inward through the disk to accrete onto the compact object. Soviet astrophysicists Nikolai I. Shakura and Rashid A. Sunyaev developed a particularly successful model of this type to explain the observational appearance of accretion onto a black hole in a close binary system (Shakura & Sunyaev 1973). British astrophysicist Geoffrey T. Bath then pointed out that unstable mass transfer from the red component of the binary system, which he had been studying, would lead to the release of energy in the accretion disk surrounding the white dwarf, and he suggested this to be the origin of the dwarf-nova outburst (Bath 1973).

The following year, Japanese astronomer Yoji Osaki suggested instead that some kind of instability within the disk itself may be responsible for the intermittent accretion of matter onto the white dwarf (Osaki 1974). In his picture, matter flows from the red star through the inner Lagrangian point at a more-or-less steady rate, collides with the matter already present in the accretion disk to form a "hot spot," which by then had been found to exist in dwarf-nova systems (Smak 1971), and gradually accumulates in the disk (see Figure 21.3). At some point, Osaki suggested, some kind of instability in the disk causes a rapid dumping of matter from the disk onto the white dwarf, which results in the dwarf nova outburst.

The nature of the disk instability was elucidated around 1980 independently by the Japanese astrophysicist Reiun Hoshi (1935–1999) and by the German astronomers Friedrich Meyer and his wife, Emmi Meyer-Hofmeister (Hoshi 1979), working at the Max-Planck Institut für Physik und Astrophysik in Munich. In studying hypothetical steady-state accretion disks, through which mass flows inward at a constant rate, they found that the inner part of the disk is fully ionized and that energy is transferred vertically away from the central plane by radiation. However, at some distance out from the central white dwarf, they found that matter in the disk is no longer completely ionized and that vertical energy transport instead occurs by convection. In the unstable intermediate region, a steady accretion flow cannot exist. Instead, as Meyer and Meyer-Hofmeister put it, "the disk alternates between short-lived radiative states of high mass flow and long-lived convective states of low mass flow..." In the convective state, mass gradually builds up in the disk until it undergoes a transition to the radiative state with high mass flow. The disk then dumps matter rapidly onto the white dwarf, as Osaki had suggested, accompanied by a substantial release of energy. When sufficient mass has been dumped from the disk, it relapses again to a low-mass-flow, convective state.

Observations soon showed that this disk-instability model for dwarf-nova systems is generally correct (Vogt 1983). Photoelectric photometry of OY Carinae showed that only the disk brightens, while the white dwarf, hot spot, and gas stream flowing from the red companion star remain relatively unaffected. A decade later, a group of investigators re-analyzed this unique set of eclipse light curves and was able to use an eclipse-mapping technique to construct from these data images of the surface of the accretion disk in quiescence and during outburst (Figure 21.5; Rutten *et al.* 1992). Confirming the earlier results, they found that the disk indeed becomes more luminous during outburst, while the white dwarf and hot spot become relatively less important. In quiescence (Figure 21.5(a)), light from the system is dominated by the white dwarf (the intense black spot at the center of this negative image). As the outburst begins, the disk rapidly becomes brighter than the white dwarf, ultimately dominating the radiant emission from the system (Figure 21.5(d)). Note the bright ring of emission around the outer edge of the disk at this stage of the eruption. During the peak of the outburst (Figures 21.5(e) and (f)), emission from the disk becomes more concentrated around the white dwarf as mass flows through the disk from the outer regions that dominate the emission in Figure 21.5(d) into the inner parts of the disk and onto the white dwarf.

Other physical effects may also be important in dwarf novae, but these topics are beyond the scope of this chapter. The interested reader is referred elsewhere for additional details (Warner 2003).

21.3 Magnetic CVs

In December 1934, British amateur astronomer J. P. M. Prentice discovered a nova in the constellation Hercules (Shapley 1934). Now called Nova DQ Herculis (1934), or simply DQ Her, it exhibits short-period fluctuations (flickering) in luminosity. In 1954 Merle F. Walker at the Mount Wilson and Palomar Observatories found that it is an eclipsing binary with the very short orbital period of $4^h 39^m$ that consists of a star with a faint, blue companion (Walker 1954). This system has the same structure

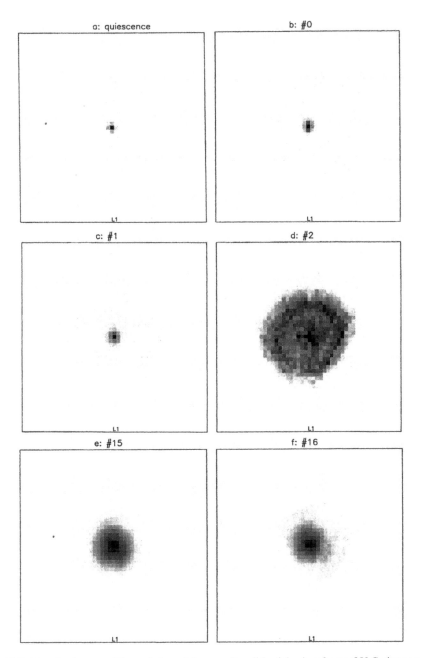

Figure 21.5. Negative images of the evolution of the accretion disk of the dwarf nova OY Carinae an during outburst. (a) The disk in quiescence. (b)–(d) Sequential stages in the development of the disk during the beginning of the outburst. (e) and (f) The accretion disk at maximum light during the outburst. From Rutten *et al.* (1992), *A&A*, **265**, 159; reproduced with permission, copyright ESO.

that we now know to be common to all the cataclysmic variables (Kraft 1958b). In further photometric observations of this system, Walker found low-amplitude variations that appeared to be periodic with a period of about 1 minute (Walker 1956); subsequent observations refined this period to 71 seconds (Walker 1974). In 1974, Geoff Bath and his collaborators interpreted this well-defined periodicity as the rotation period of a white dwarf with a magnetic field of a million gauss or more (megagauss or MG)[2] that is offset from the rotation axis (Bath *et al.* 1974). Accretion onto the magnetic poles produces hot spots, which rotate in and out of the observer's line of sight with that period. A magnetic field of the inferred size also is sufficient to disrupt the some of the hot, ionized inner regions of the accretion disk formed by mass transfer in such a close binary system (Figure 21.6). The most significant difference between this figure and Figure 21.3 is that in Figure 21.6 the strong magnetic field of the white dwarf (in the center of the accretion disk in this sketch) has completely disrupted the inner part of the accretion disk around it. Consequently, mass flows through the residual outer part of the disk and then along the magnetic-field lines, which funnel it down onto the magnetic polar caps of the white dwarf.

DQ Herculis thus became the prototype of a new class of magnetic CVs, which are now called "intermediate polars." At the beginning of the 21st century some 26 CVs of this type were known (Wickramasinghe & Ferrario 2002).

In 1977, another object was found that proved to have an even stronger field. Richard A. Berg and J. Graeme Duthie (1934–2019), working at the University of

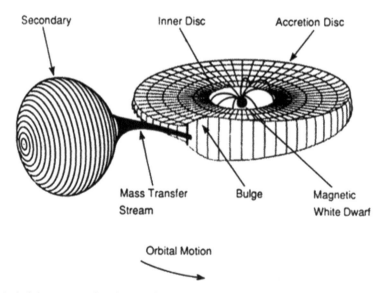

Figure 21.6. Artist's concept of an intermediate polar like the magnetic CV DQ Herculis. Source: http:// heasarc.gsfc.nasa.gov/docs/objects/cvs/cvstext.html.

---

[2] For comparison, the field strength at Earth's magnetic poles is less than 0.75 gauss.

Rochester, carried out high-speed photometry of the variable AM Herculis and found that it exhibits irregular variations on timescales of minutes but with no stable pulsations (Berg & Duthie 1977). They recognized its similarity to cataclysmic variables, and they pointed out that it lay within the error box established by SAS-3 X-ray satellite observations for the X-ray source 3U 1809+50 (Hearn *et al.* 1976). They also noted that the variability of the X-ray source was similar to the variability they found in AM Her. Further observations showed that this unusual object is an eclipsing binary with an orbital period of about 3.1 hours, seen in both the X-ray and visible bands, and linear and circular polarization revealed that it has a magnetic field of about 200 MG (later found to be overestimated by about a factor of 10; Cowley *et al.* 1977). This field is strong enough to prevent the formation of an accretion disk. Instead, in this system the magnetic white dwarf is rotating synchronously with the orbital motion; i.e., like the Earth and Moon, it always keeps the same face to its companion. The stream of ionized gas from the companion is then channeled by the magnetic field directly onto the white dwarf, forming an X-ray-emitting hot spot above the near magnetic pole (Chanmugam *et al.* 1977).

A quarter century later, about 60 systems of this type were known—now called "polars"—about 20% of all CVs (Wickramasinghe & Ferrario 2002). The synchronous rotation of the two components is maintained by the strong magnetostatic torques between the two stars. The X-ray emission originates in the accretion shocks, while the optical emission is due to cyclotron radiation from the shocks, photospheric emission from the white dwarf, and emission from the accretion stream, the accretion funnel above the magnetic polar cap, and the companion star. In three of the 60 AM Her systems, the white dwarf rotates slightly out of synchronism, possibly due to recent nova eruptions that have temporarily disrupted the system. The magnetic-field strengths of most AM Hers lie in the range from about 7 MG to 80 MG, although one (AR U Ma) has a field strength of about 230 MG.

21.4 AM CVn Stars

In 1947, Milton Humason (1891–1972) and Fritz Zwicky (1898–1974) were using the 18-inch Schmidt telescope at the Mount Palomar Observatory to search for faint, blue stars that might be white dwarfs. In the Hyades cluster they found 14 white-dwarf candidates, and in the region of the north galactic pole they discovered 31 more. To confirm that a candidate star is indeed a white dwarf requires an examination of its spectrum, so a decade later Jesse L. Greenstein (1909–2002) and his collaborators began using the 200-inch (5-meter) Hale telescope at Mount Palomar to obtain spectra of hundreds of faint, blue stars from various candidate lists. Among them were the candidates identified by Humason and Zwicky. In particular, Greenstein and Mildred Shapley Matthews (1915–2016), the daughter of Harvard astronomer Harlow Shapley (1885–1972) and his wife, examined the spectra of several white-dwarf candidates with unusual spectroscopic features, among them HZ 29, the 29th object in Humason and Zwicky's candidate list (Greenstein & Matthews 1957). They noted that "HZ 29 has the most peculiar spectrum observed." They found no evidence of hydrogen, and most of the strong features in the spectrum matched well with the lines of neutral helium (He I). However, the spectrum differed substantially from a normal helium-rich white dwarf, with the lines in HZ 29 being extremely shallow and diffuse. Indeed, the lines appeared almost double, suggesting rotation. In addition, the photometric colors indicated that it is an extremely hot object.

A decade later, Polish astronomer Józef Smak found HZ 29 to be a variable star with a period of about 18 minutes, and he postulated that it might actually be a very close binary system (Smak 1967). High-speed photometric observations of this object refined the period to 1051 seconds (Ostriker & Hesser 1968), which Brian Warner and Edward L. ("Rob") Robinson interpreted as the orbital period of a very close binary system. They accordingly proposed that it is a late stage in the evolution of a cataclysmic variable from which all the residual hydrogen has been removed. In 1972, John Faulkner, Brian P. Flannery, and Warner offered a more detailed interpretation of this system, which they noted would henceforth be known as AM Canum Venaticorum (AM CVn) to variable-star observers (Faulkner *et al.* 1972). In their model, the mass-accreting primary is an ordinary helium (DB) white dwarf, while the companion star that fills its Roche lobe and transfers mass to the white dwarf is itself a degenerate, very-low-mass He white dwarf, with a mass of about 0.04 M_{Sun}. The flickering observed in the visible band provides evidence for mass transfer through the inner Lagrangian point to a hot spot on a ring surrounding the white-dwarf primary, just as in an ordinary CV. The significant difference, however, is that the mass being transferred in HZ 29 (= AM CVn) is He instead of hydrogen.

Such ultrashort-period binaries also are of interest because of the role gravitational radiation plays in the evolution of the system (Faulkner 1971). By 1995, six of these types of systems were known, and Warner has suggested that they may be the strongest sources of gravitational radiation in the 0.7–2 mHz range (Warner 1995). Both the evolutionary origin and the evolutionary fate of these systems remain of interest but are beyond the scope of this chapter. The reader interested in additional details is referred to the literature and references therein (Brooks *et al.* 2015).

References

Bath, G. T. 1969, ApJ, 158 571

Bath, G. T. 1972, ApJ, 173, 121

Bath, G. T. 1973, NPhS, 246, 84

Bath, G. T., Evans, W. D., & Pringle, J. E. 1974, MNRAS, 166, 113

Berg, R. A., & Duthie, J. G. 1977, ApJ, 211, 859

Brooks, J., Bildsten, L., Marchant, P., & Paxton, B. 2015, ApJ, 807 74; see also
    Ramsay, G., et al. 2018, A&A, 620, A141
      Green, M., et al. 2019, www.astrosen.unam.mx/CWDB2019/GreenM_cwdb2019.pdf

Casanova, J., et al. 2011, Natur, 478, 490

Chanmugam, G., & Wagner, R. L. 1977, ApJ, 213 L13; see also
    Stockman, H. S., Schmidt, G. D., Angel, J. R. P., Liebert, J., Tapia, S., & Beaver, E. A. 1977, ApJ, 217, 815

Cowley, A. P., & Crampton, D. 1977, ApJ, 212 L121; see also
    Tapia, S. 1977, ApJ, 212, L125
    Hearn, D. R., & Richardson, J. A. 1977, ApJ, 213, L115
Crawford, J. A., & Kraft, R. P. 1956, ApJ, 123, 44
Faulkner, J. 1971, ApJ, 170, L99
Faulkner, J., Flannery, B. P., & Warner, B. 1972, ApJ, 175, L79
Gehrz, R. D., et al. 1998, PASP, 110, 3
Glasner, S. A., & Truran, J. W. 2012, JPhCS, 337, 012071
Greenstein, J. L., & Kraft, R. P. 1959, ApJ, 130, 99
Greenstein, J. L., & Matthews, M. S. 1957, ApJ, 126, 14
Hearn, D. R., Richardson, J. A., & Clark, G. W. 1976, ApJ, 210, L23
Hind, J. R. 1856, MNRAS, 16, 56
Hoshi, R. 1979, PThPh, 61 1307; see also
    Meyer, F., & Meyer–Hofmeister, E. 1981, A&A, 104, L10
Joy, A. H. 1954, ApJ, 120, 377
Kippenhahn, R., & Thomas, H. C. 1978, A&A, 63 265; see also
    Sparks, W. M., & Kutter, G. S. 1987, ApJ, 321, 394
Kraft, R. P. 1958a, ApJ, 127, 625
Kraft, R. P. 1958b, ApJ, 130, 110
Kraft, R. P. 1964, ApJ, 139, 457
Kraft, R. P. 2009, ARA&A, 47, 1
Krzeminski, W. 1965, ApJ, 142, 1051
Kutter, G. S., & Sparks, W. M. 1972, ApJ, 175 407; see also
    Starrfield, S., et al. ApJ, 176, 169
    Starrfield, S., Sparks, W. M., & Truran, J. W. 1974, ApJS, 28, 247
McLaughlin, D. B. 1950, PA, 58, 50
McLaughlin, D. B. 1943, POMic, 8 149; see also
    McLaughlin, D. B. 1960, in Stellar Atmospheres, Greenstein, J. L. (ed.) (Chicago, IL: Univ. Chicago Press) 585
McLeod, N. W. 1948, ASPL, 5, 265
Osaki, Y. 1974, PASJ, 26, 429
Ostriker, J. P., & Hesser, J. E. 1968, ApJ, 153 L151; see also
    Warner, B., & Robinson, E. L. 1972, MNRAS, 159, 101
Paczyński, B. 1965, AcA, 15, 197
Rutten, R. G. M., et al. 1992, A&A, 265, 159
Saslaw, W. C. 1968, MNRAS, 138, 337
Schatzman, E. 1949, AnAp, 12, 281; see also
    Eminadze, T.A. 1962, SvA, 6, 431
Schatzman, E. 1951, AnAp, 14, 294
Shakura, N. I., & Sunyaev, R. A. 1973, A&A, 24, 337
Shapley, H. 1934, HarAC, 318,
Smak, J. 1967, AcA, 17, 255
Smak, J. 1971, AcA, 21 15; see also
    Warner, B., & Nather, R. E. 1971, MNRAS, 152, 219
Starrfield, S., et al. 1998, MNRAS, 296, 502
Starrfield, S., Iliadis, C., & Hix, W. R. 2016, PASP, 128, 051001

Stephenson, F. R. 1976, QJRAS, 17, 121

Unknown, 1896, Natur, 55, 84

Vogt, N. 1983, A&A, 128, 29

Walker, M. F. 1954, PASP, 66, 230

Walker, M. F. 1956, ApJ, 123, 68

Walker, M. F. 1974, ApJ, 134, 171

Warner, B. 1995, Ap&SS, 225, 249

Warner, B. 2003, Cataclysmic Variable Stars (Cambridge: Cambridge Univ. Press); see also
    Hameury, J. M. 2020, AdSpR, 66, 1004

Wickramasinghe, D., & Ferrario, L. 2002, ASP Conf. Ser., Vol. 261 (San Francisco, CA: ASP) 82

Chapter 22

The First Stars

The night sky today is spangled with stars, but in the beginning there were no stars. How and when the *first stars* appeared—and what their physical properties were—is a fascinating topic in modern cosmology.[1] At the time of this writing, we are only beginning to learn about them.

The best scientific evidence indicates that about 13.7 billion years ago the universe began the expansion that is still going on today (see Figure 22.1; Haemmerlé *et al.* 2020). Initially, the temperature and density were unimaginably large, and after the first several minutes the universe consisted solely of dark matter and a fully ionized mixture of hydrogen, helium, and a smattering of the light element Li, but no heavier elements (see Chapter 17).

After about 400,000 years, the universe first became cool enough for electrons to combine with the positively charged atomic nuclei to produce neutral atoms. This is called the "Epoch of Recombination," even though never before in the short history of the universe to that point had electrons and ions been combined into atoms. As Steven Weinberg put it (Weinberg 1977), "Just before the electrons started to be captured into atoms, at a temperature of about 3000 K, the pressure of radiation was enormous, and the Jeans mass was correspondingly large, about a million or so times larger than the mass of a large galaxy [itself a hundred billion times the mass of the Sun].... However, a little later the electrons joined with nuclei into atoms; with the disappearance of free electrons, the universe became transparent to radiation; and so the radiation pressure became ineffective... The Jeans mass[2] dropped... to about one-millionth the mass of a galaxy."

After recombination, the universe continued to expand and cool, and as the temperature declined, the intensity of the background radiation declined as well,

---

[1] Cosmology is the science of the origin and evolution of the universe.

[2] The Jeans mass is the mass of gas at which the inward-directed gravitational attraction is exactly balanced by outward-directed pressure forces (see Chapter 11). For objects with masses larger than this, the gravitational force is stronger, and the object undergoes collapse.

22-1

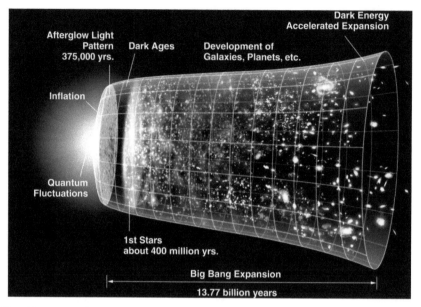

Figure 22.1. Illustration of major stages early in the expansion of the universe. Source: upload.wikimedia.org/wikimedia/commons/6/6f/CMB_Timeline300_no_WMAP.jpg.

becoming fainter and fainter and redder and redder. The universe entered what has been called the "Cosmic Dark Ages." The stage was set for the appearance of the first stars.

22.1 Formation of the First Stars[3]

The first structures to form in the early universe were concentrations of dark matter —which astronomers call "halos"—containing more than a million times as much mass as our own Sun (Norman *et al.* 1999; Bromm 2013). The gravitational attraction of the dark matter halos caused ordinary baryonic gas—made up of the elements formed during the first few minutes in the expansion of the universe—to flow into the centers of these halos (Miralda-Escudé 1999). These concentrations of baryonic gas were the stellar "nurseries" in which the first stars formed (Larson 1999). As gas flowed into the centers of these halos, it became compressed and heated. During star formation today, the presence of dust grains formed from heavier elements like carbon, oxygen, silicon, etc., radiates this heat away efficiently, allowing the collapse to continue and form stars. However, the matter that formed the first stars—called "Population III stars" by astronomers—contained no heavy elements at all, so there were no dust grains. Instead, heat was radiated away by molecular hydrogen (H_2), a process that was first recognized to be important in the late 1960s (Saslaw & Zipoy 1967). However, the minimum temperature to which H_2

---

[3] At the time of this writing, the launch of the James Webb Space Telescope in December 2021 is beginning to revolutionize our understanding of the first stars and galaxies. Consequently, some of the material discussed in this final chapter may be outdated by the time this book is published.

can cool the gas is about 200 K (Bromm 2013), about 10 times larger than the minimum temperature in star-forming regions today. Consequently, the Jeans mass was much larger in the dark matter halos.

Results from simulations carried out around the turn of the millennium (Norman et al. 1999; Bromm et al. 2002), showed that, generally, "the collapsing region makes a central massive clump with a typical mass of hundreds of solar masses... Soon after its formation, the clump becomes gravitationally unstable and undergoes runaway collapse at a roughly constant temperature... The central clump does not typically undergo further subfragmentation and is expected to form a single star" (Loeb 2010). The masses of the first stars may actually have extended over a wide range from about 1 M_{Sun} up to an upper mass limit variously estimated to be of order 100 M_{Sun} to 500–600 M_{Sun} or more (Bromm 2013). However, the angular momentum of the collapsing gas causes it to form an accretion disk around the central object and then to undergo fragmentation, producing a compact cluster of stars with a wide range of masses, rather than a single high-mass star (Clark et al. 2011; Haemmerlé et al. 2020). This is currently understood to be the standard pathway for the production of the first stars.

Quite different kinds of objects were formed depending upon the rate of accretion onto the protostellar cores (Clark et al. 2011; Haemmerlé et al. 2020). For accretion rates less than about 10^{-3} M_{Sun} per year, the resulting first stars are thought to have been massive (of the order of 100 M_{Sun}), compact, and very hot objects (see below), which thus radiate strongly in the ultraviolet. As one group of investigators put it, "In popular cosmological scenarios, massive stars within early protogalaxies, aided perhaps by a population of accreting seed black holes, generated the UV and x-ray radiation that reheated and reionized the universe" (Haemmerlé et al. 2020). On the other hand, for accretion rates greater than about 10^{-2} M_{Sun} per year, the forming star remained bloated and "fluffy." It thus did not radiate much UV, and this may have allowed accretion onto the central object to continue for a long time, ultimately forming a supermassive star (see below).

22.2 Unique Properties of the First Stars

Because the Big Bang produced no elements heavier than lithium (Li)—and in particular, no carbon, nitrogen, or oxygen—the first stars could not be powered by the CNO reactions of the Bethe cycle. Instead, the energy they produced initially came solely from gravitational contraction and from hydrogen- (and deuterium-) burning nuclear reactions. Because the rates of these reactions increase much more slowly with the central temperature than do the CNO reactions, the first stars had to contract to appreciably higher central temperatures than current generations of stars in order to generate enough energy to stabilize at a constant luminosity. Eddington had already pointed out early in the 20th century that higher central temperatures result from contraction to smaller radii, which in turn implies higher effective ("surface") temperatures. Accordingly, for a given luminosity, the first stars were significantly hotter—and thus bluer—than stars in the sky today.

The absence of heavy elements in the composition of the primordial gas also meant that the opacity of stellar matter was greatly reduced compared to stars that exist today. In stars with compositions similar to that of the Sun, the continuum opacity throughout much of the stellar interior is provided by free-free or bound-free electronic transitions in iron and other heavy elements, all of which were absent in the first stars. The low opacity also meant that radiation was able to transport energy efficiently throughout the stellar interior. Consequently, the thermal structure of the first stars was quite different from those that exist today.

Selected properties of zero-age Main Sequence (ZAMS) models for the first stars are tabulated in Appendix G. Comparing the 60 M_{Sun} model from Appendix C, which has element abundances like those of the Sun, with the 60 M_{Sun} model from Appendix G illustrates these differences. The radius of the model with solar composition is 10.2 R_{Sun}, while the model with no heavy elements has a radius of only 3.1 R_{Sun}. The luminosities of the two models are very similar, and because of the much smaller radius, the effective temperature of the model without heavy elements is 87,700 K, as opposed to 48,195 K for the model with heavy elements.

Because the internal temperatures in the first stars were so high, the hydrogen and helium gas of which they were composed was fully ionized throughout, and radiation pressure completely dominated over the gas pressure. Eddington showed in 1926 that, as the luminosity of a star increases, there comes a point at which its self-gravity can no longer hold the star together against the force exerted by the radiation pressure. This limit—now called the "Eddington luminosity" in his honor —is proportional to the stellar mass. Models with smaller luminosities are stable, while a model with a higher luminosity would experience rapid mass loss. An approximate expression for the Eddington luminosity is (Hansen & Kawaler 1994) $L_{Edd}/L_{Sun} \approx 3.4 \times 10^4 \, (M/M_{Sun})$. As the table in Appendix G shows, models for the first stars that have masses in excess of about 150 M_{Sun} have luminosities that exceed about half of the Eddington luminosity.

22.3 Evolution of the First Stars

Evolutionary calculations (Marigo *et al.* 2001) show that—because they have to contract to very high central temperatures[4] in order for the nuclear reactions of the *p–p* chain to provide sufficient energy to sustain the stellar luminosity—the massive first stars reached temperatures high enough to ignite the 3-α reaction while they were still contracting toward the hydrogen-burning Main Sequence. This is utterly unlike the situation with stars in the sky today. As Marigo *et al.* put it, He-ignition "marks a fundamental event, as it represents the first significant production of metals in these stars. As a consequence, the synthesis of primary ^{12}C [by these helium-burning reactions] leads to the activation of the CNO-cycle, which then starts providing nuclear energy in competition with the *p–p* reactions.... At a sufficiently high mass ($M \geqslant 20$ M_{Sun} in our models), the 3-α reaction ignites even *before* the *p–p* reactions have slowed down the initial stellar contraction. In these models then, H-burning simply proceeds via the CNO-cycle, without any significant phase of central burning via the *p–p* chain."

---

[4] The central temperatures of the ZAMS models for the massive first stars listed in Appendix G significantly exceed 10^8 K, high enough to ignite He-burning.

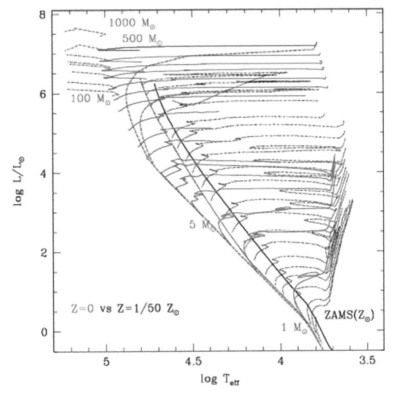

Figure 22.2. Evolutionary tracks of the first stars in the Hertzsprung–Russell diagram. Credit: Schaerer, D. 2002, A&A, **382**, 28, reproduced with permission; copyright ESO. See text for details.

Figure 22.2 shows evolutionary tracks for Population III stars in the Hertzsprung–Russell diagram (Schaerer 2002). The solid black line in the figure is the zero-age Main Sequence (ZAMS) for models with heavy-element abundances equal to those in the Sun. The solid gray line to the left (i.e., at somewhat higher effective temperatures) is the ZAMS for models with heavy-element abundances equal to one-fiftieth of the solar abundances, and the dotted gray lines represent evolutionary tracks of models with a variety of masses for the same composition. The solid blue lines are the evolutionary tracks for Population III stellar models, which have no heavy elements. The left-hand (high-T_{eff}) ends of these tracks trace out the Population III ZAMS. It extends to much higher effective temperatures than does the corresponding ZAMS for stars with even small amounts of heavy elements because the Population III stars contract to much higher central temperatures. The two dotted red curves that extend upward along the ZAMS and then curve to the right (toward lower effective temperatures) represent isochrones[5] at 2 million years

---

[5] An isochrone is a plot at a given elapsed time of the locations in the Hertzsprung–Russell diagram of stars with a broad range of masses that all started out on the ZAMS at the same time. Because stars with different masses evolve at very different rates, lower-mass stars may still be on the Main Sequence at a time when higher-mass stars have already evolved to the red-giant branch or beyond.

(the upper curve) and 4 million years. Comparison of the uppermost, blue evolutionary track with the uppermost red isochrone shows that in less than 2 million years this 1000 M_{Sun} model has completed its H- and He-burning phases of evolution, its outer envelope has expanded while its carbon–oxygen core is continuing to contract, and the central temperature is on the verge of igniting carbon-burning nuclear reactions. Even the much-less-massive 100 M_{Sun} model runs through the same phases in less than 4 million years. In general, the H- and He-burning lifetimes of massive Population III stars are about 2×10^6 years and 2×10^5 years, respectively, almost independent of the mass, for models with masses between 120 and 1000 M_{Sun} (Marigo *et al.* 2003).

The internal structures of these massive first stars also are quite different from those of stars with significant heavy-element abundances. This is illustrated in Figure 22.3 for a 250 M_{Sun} model calculated by Marigo *et al.* (2003). Because of the very high rate of energy generation in the core of this massive star, an extensive convection zone develops covering the innermost 90% of the star's mass, as shown by the upward-slanting cross-hatching in the figure. Temperatures are high enough for significant nuclear burning through more than half of the stellar interior (the downward-slanting cross-hatching). Central H burning is essentially complete after about 2.2 million years, by which time the star has already moved well to the right in

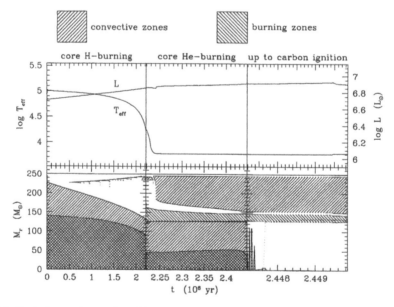

Figure 22.3. Evolutionary changes for a 250 M_{Sun} Population III star. Credit: Marigo, P., *et al.* 2003, A&A, **399**, 617, reproduced with permission; copyright ESO. The horizontal axis at the bottom of this figure shows the age of the star in millions of years. The upper panels show changes in the luminosity of the star (on the right-hand vertical axis) and in its effective temperature (on the left-hand vertical axis). The major nuclear-burning stages are labelled across the top of the figure. The bottom panels are Kippenhahn diagrams that show the extent in mass (represented along the left-hand vertical axis) of the nuclear-burning zones (downward-slanting cross-hatching) and of the convective regions (upward-slanting cross-hatching) in this stellar model.

the Hertzsprung–Russell diagram. At this point, the scale of time on the horizontal axis of Figure 22.3 changes. After this point, H burning continues in a thick shell just outside the large convective core, while He burning continues in about the central 20% of the stellar mass. In just another 40,000 years, a deep subsurface convection zone develops, extending over the outermost 30% of the star's mass, as the star reaches the Hayashi track. He burning is essentially complete about 450,000 years after the start of the shell-burning phase, and in less than 2500 years more, the central temperature has reached the point of carbon ignition, and the calculation ends.

22.4 Deaths of the First Stars

Every star evolves toward progressively higher central temperatures and densities, pausing for varying lengths of time as periods of nuclear burning liberate enough energy to halt the continuing, relentless gravitational contraction temporarily. This is as true for the first stars as for stars of all later generations. While the core of a star continues to contract, its outer layers execute a complicated dance—sometimes distending outward to enormous radii, sometimes contracting back in to become little more than a thin veneer covering the core—in response to changes in its internal structure. During some of these phases, a star may lose mass of greater or lesser amounts. Figure 22.4 shows the final remnant masses for zero-metallicity (Population III) stars as a function of their initial masses (Heger & Woosley 2002). As this figure shows, if a star ends up with a mass less than the Chandrasekhar limit (about 1.4 times the mass of our Sun), the electrons inside it become degenerate, preventing further contraction, and the star settles down to become a white dwarf. Above this limit, core contraction proceeds until the star either explodes as a supernova or implodes directly to form a black hole (Marigo *et al.* 2003; Joggerst & Whalen 2011). If the initial mass is greater than about 6 to 8 times the solar mass but less than about 26 M_{Sun}, a Population III star's life ends in a Type II supernova explosion. "Unlike lower-mass stars… no point is ever reached at which a massive star can be fully supported by electron degeneracy. Instead the center evolves to ever higher temperatures, fusing ever heavier elements until a core of iron is produced. The collapse of this iron core to a neutron star releases an enormous amount of energy, a tiny fraction of which is sufficient to explode the star as a supernova…. The calculated neutron star masses, supernova light curves, and spectra from these models are … consistent with observations… Such stars are capable of producing, with few exceptions, the isotopes between mass 16 and 88 as well as a large fraction of still heavier elements by the *r* and *p* processes" (Woosley *et al.* 2002).

Above this initial mass but below about 40 M_{Sun}, a supernova explosion releases some 10^{51} ergs of energy, disrupting the star following a final burst of explosive nuclear burning but leaving a black hole remnant (Greif 2015). For Pop III stars with initial masses between about 40 and 100 M_{Sun}, the entire star collapses directly into a black hole. Between about 100 and 140 M_{Sun}, the star is subject to pulsational instability that leads to mass loss, but the star still leaves a black hole remnant. Population III stars with initial masses between 140 and 260 M_{Sun} become subject to

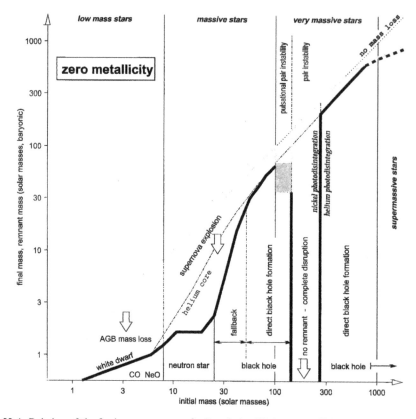

Figure 22.4. Relation of the final remnant mass of a Population III (zero-metallicity) star to its initial mass. From https://2sn.org, after Heger, A., and Woosley, S. E. (2002). See text for details.

the pair instability (Barkat *et al.* 1967). This occurs when the temperature becomes high enough for thermal photons to create electron–positron pairs. The process drains energy from the radiation field, weakening the radiation-pressure support for the star. This leads to collapse "and the ignition of explosive nuclear burning. The subsequent thermonuclear runaway reverses the collapse and ejects the entire star, leaving no remnant behind" (Pan *et al.* 2012). These stars "are completely disrupted with explosion energies of up to ~10^{53} erg and eject up to ~60 M_{Sun} of ^{56}Ni" (Heger *et al.* 2001). Finally, stars with masses in excess of 260 times the solar mass collapse directly to become black holes at the ends of their nuclear-burning lifetimes.

22.5 Supermassive Pop III Stars: Seeds of the SMBHs in the Centers of Galaxies

The first hint that the centers of galaxies might contain what we now call supermassive black holes (SMBHs)—with masses millions to billions of times larger than the mass of the Sun—originated with the Armenian–Soviet astronomer Viktor A. Ambartsumian (1908–1996). At the 11th Solvay conference in 1958, he summarized the outcome of his observational studies of external galaxies with the statement, "galactic nuclei must contain bodies of huge mass and unknown nature" (Israelian 1997). Also in the 1950s, during surveys of the universe at radio wavelengths, astronomers from Cambridge University discovered some strong sources of radio emission with very small angular sizes, which they accordingly termed "quasi-stellar" radio sources, soon shortened to "quasars." By 1963, several quasars had been identified with star-like objects (Matthews & Sandage 1963). That same year, Caltech astronomer Maarten Schmidt found that the mysterious emission lines in the quasar spectra—which until then astronomers had not been able to identify with any known chemical elements—were actually the Balmer lines of H and an emission line from Mg and one from O, all redshifted to $z = 0.158$. This put the source, 3C 273, at a cosmological distance, implying that it is emitting energy at a truly phenomenal rate. Going back over previously recorded images that include this source, astronomers also found that the light from 3C 273 varies irregularly on timescales of a year or so, indicating that a significant fraction of the light it emits originates from a region less than about a lightyear in diameter (Shields 1999). Taken together, these facts suggested that the enormous emission of radiant energy is generated by matter falling into an SMBH (Salpeter 1964; Zel'dovich 1964). However, the "growing acceptance of the black hole model [for active galactic nuclei (AGNs)] resulted, not from any one compelling piece of evidence, but rather from the accumulation of observational and theoretical arguments suggestive of black holes and from the lack of viable alternatives…" (Shields 1999)

The first actual image of an SMBH in an AGN—or, more accurately, of the region immediately around it—was obtained only recently, when the "Event Horizon Telescope" (EHT) obtained the first image of the supermassive black hole M87* in the active nucleus of the galaxy Messier 87 (The Event Horizon Telescope Collaboration 2019). The EHT is a global consortium of eight sensitive radio telescopes, and the EHT collaboration was awarded the 2020 Breakthrough Prize in Fundamental Physics for this achievement (Feinberg 2019).

Evidence accumulated over the past 70 years suggests that SMBHs exist in normal galaxies as well. There is even a massive black hole in the center of our own Milky Way Galaxy; it is known as the active radio source Sagittarius A*. Astronomers Reinhard Genzel and Andrea Ghez shared half of the Nobel Prize in 2020 for proving observationally that Sgr A* is a black hole with a mass about four million times the mass of our Sun (Berkowitz 2020). The EHT collaboration has also obtained an image of Sgr A* (The Event Horizon Telescope Collaboration 2022). Because the mass of M87* is about a thousand times greater than that of Sgr A*, and because the size of a black hole scales with its mass, the physical dimensions of M87* are vastly larger than are those of Sgr A* (see Figure 22.5). The intensities in these images correspond to the intensity of the radio emission. The black holes are the low-intensity regions ("shadows") in the centers of the images, and the bright rings are produced by radiation bent around the black holes by their strong gravitational fields and perhaps by emission from matter (e.g., stars) accreting into the SMBHs.

Recent cosmological simulations have shown that "the rare convergence of strong, cold accretion flows" (Latif *et al.* 2022) can form supermassive stars

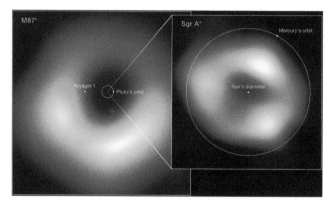

Figure 22.5. Comparison of the images of the supermassive black hole M87* in the center of the active galaxy Messier 87 and the SMBH Sgr A* in the center of the Milky Way Galaxy, both obtained with the Event Horizon Telescope. Various scales in the solar system are shown for comparison. Credit: EHT collaboration (acknowledgment: Lia Medeiros, xkcd); source: https://eso.org/public/images/eso2208-eht-mwe/.

(SMSs) that grow in mass to tens of thousands of solar masses before undergoing direct collapse into black holes. In these simulations, the primordial (baryon plus dark matter) halo reaches "a mass of $3.9 \times 10^5 \, M_{Sun}$ at $z = 35$, the approximate minimum mass at which it could form Population III ... stars. However, highly supersonic turbulence driven by the cold flows from large scales prevent gas in the halo from collapsing into stars. ... Collapse is instead triggered when infall velocities finally exceed turbulent velocities... A dense clump appears in the core 5 kyr after the onset of collapse... However, before this clump can form a star it becomes part of a much larger clump that forms a supermassive star" (Latif *et al.* 2022).

Two massive clumps—which Latif *et al.* label C1 and C2 and which become supermassive stars SMS1 and SMS2, respectively—form within the first 100,000 years (100 kyr). "C1 initially grows mainly through accretion but merges with other clumps 200 kyr later... [These] collisions raise its mass to $10^4 \, M_{Sun}$ over 400 kyr and it then doubles in mass over the next 600 kyr... C1 reaches a mass of $2.6 \times 10^4 \, M_{Sun}$ just before merging with another cluster of clumps at the end of the run." See panel (a) in Figure 22.6.

In contrast, "C2 grows to 1000 M_{Sun} in 100 kyr before multiple encounters with dense filaments increases its mass by a factor of four over the next 100 kyr. It then grows to 8000 M_{Sun} in about 900 kyr primarily by accretion... C2 merges with a few... clumps and grows by a factor of five in about 200 kyr... C2's final mass at the end of the simulation is $4.1 \times 10^4 \, M_{Sun}$." See panel (b) in Figure 22.6. Both supermassive stars reach effective temperatures of 100,000–120,000 K before collapsing directly into black holes at masses of 31,316 M_{Sun} and 39,765 M_{Sun}. respectively. Thus, "Cold flows drive violent, supersonic turbulence in the [dark matter] halo that prevents star formation until it reaches a mass that triggers sudden, catastrophic baryon collapse that forms 31,000 and 40,000 solar-mass stars ... [ensuring] that haloes capable of forming quasars by $z > 6$ produce massive seeds" (Latif *et al.* 2022).

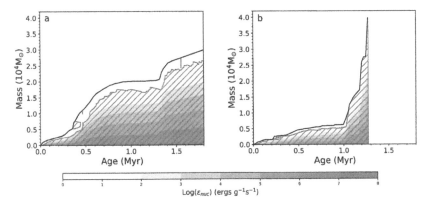

Figure 22.6. Growth in mass of two supermassive Population III stars in the early universe. See text for details. (a) Clump C1 (SMS1) and (b) clump C2 (SMS2). From Latif, M. A., *et al.* (2022), Extended Data Figure 6; reproduced with permission from Springer Nature under license 5504931097557. The vertical axis in each panel shows the mass in units of 10^4 M_{Sun}, and the horizontal axis gives the age in millions of years (Myr). The black line represents the surface of the star, the red hatched regions represent convection zones, and the blue shading represents thermonuclear energy generation corresponding to the color bar at the bottom of the figure. For comparison, the thermonuclear energy-generation rate in the center of the present Sun is $\varepsilon_{nuc} = 17.5$ ergs g^{-1} s^{-1}. That is, the energy-generation rate in the centers of these supermassive stars is about 10 million times greater than the rate at the center of the Sun.

22.6 The First Stars in the Early Universe

The births of the first stars brought light into the early universe, ending the Cosmic Dark Ages. These very hot and luminous stars produced copious amounts of ultraviolet radiation. The UV photons ejected electrons that had been locked up in neutral hydrogen atoms since the so-called Epoch of Recombination, about 400,000 years after the Big Bang. This marked the beginning of the Epoch of Re-ionization that began a few hundred million years later.

As I write the final chapter of this book, relevant early science results have begun to be reported by astronomers using the newly commissioned James Webb Space Telescope (JWST; https://en.wikipedia.org/wiki/James_Webb_Space_Telescope# From_launch_through_commissioining). Launched on Christmas Day in 2021, the JWST took about a month to travel to its parking destination at the L2 Lagrange point, about 1.5 million kilometers on the opposite side of Earth from the Sun. Its 18 gold-plated mirror segments were successfully deployed and positioned *en route* to the L2 point, where it arrived on 2022 January 24. The near-infrared camera (NIRCam) received first light on 2022 February 3, and the first three instrument-alignment phases were completed before the end of the month. The first images from JWST were released publicly on 2022 July 11.

In two Early Release Science deep fields, Rohan Naidu and his colleagues found two very-high-redshift primordial galaxies: GLASS-z11 and GLASS-z13, with cosmological redshifts estimated from photometry to be $z \approx 11$ and 13, respectively (Naidu *et al.* 2022). At these high redshifts, the more-distant object formed about 340 million years after the Big Bang, and Naidu *et al.* estimate that it contained about a billion solar masses within a radius of about 500 pc. Similarly, the nearer galaxy formed about 430 million years after the Big Bang and contained about 2.5 billion solar masses within about 700 pc. These masses and dimensions are comparable to those of the small and relatively nearby elliptical galaxies NGC 147, 185, and 221 (also known as M32 in the constellation Andromeda). While these JWST results are still very preliminary, and they do not yet tell us very much about the first stars, the very existence of these small galaxies demonstrates that the first stars did form within a few hundred million years after the Big Bang. As astronomers continue to exploit the capabilities of the JWST, we can expect to learn much more about the actual properties of the first stars to form very early in the evolution of the universe.

The deaths of the first stars made additional contributions that would affect the later, still-unfolding history of the universe. These stars were the furnaces within which the first heavy elements in the universe were forged—carbon, nitrogen, oxygen, silicon, iron, and so on. The deaths of the first stars in violent supernova explosions expelled these heavy elements out into space. There they mixed with the primordial hydrogen and helium gas and later became incorporated into subsequent generations of stars. In generation after generation, the abundances of the heavy elements continued to increase. Eventually the interstellar gas accumulated sufficient heavy elements so that particles of ices and rock and metal were able to grow into ever-larger solid bodies. The stage was set for the formation of planets and living things.

None of this would have occurred without the first stars.

References

Barkat, Z., Rakavy, G., & Sack, N. 1967, PhRvL, 18 379; see also
    Rakavy, G., & Shaviv, G. 1967, ApJ, 148, 803

Berkowitz, R. 2020, PhT, 73, 17

Bromm, V. 2013, RPPh, 76, 112901

Bromm, V., Coppi, P. S., & Larson, R. B. 2002, ApJ, 564, 23

Clark, P. C., et al. 2011, Sci., 331 1040

Feinberg, R. T. 2019, https://aas.org/posts/news/2019/09/event-horizon-telescope-team-receives-
    breakthrough-prize; see also
    Clery, D. 2019, Sci., 366, 1434

Greif, T. H. 2015, ComAC, 2, 3

Haemmerlé, L., et al. 2020, SSRv, 216, 48

Hansen, C. J., and Kawaler, S. D. 1994, Stellar Interiors (New York: Springer) 74; see also
    Loeb, 2010, How Did the First Stars and Galaxies Form? (Princeton, NJ: Princeton Univ.
    Press) 74

Heger, A., Baraffe, I., Fryer, C. L., & Woosley, S. E. 2001, NuPhA, 688, 197

Heger, A., & Woosley, S. E. 2002, ApJ, 567, 532

https://en.wikipedia.org/wiki/James_Webb_Space_Telescope#From_launch_through_commissioining

Israelian, G. 1997, BAAS, 29, 1466

Joggerst, C. C., & Whalen, D. J. 2011, ApJ, 728, 129

Larson, R. B. 1999, in The First Stars, Weiss, A., Abel, T. G., & Hill, V. (eds.) (Berlin: Springer-Verlag) 343

Latif, M. A., et al. 2022, Natur, 607, 48

Loeb, 2010, How Did the First Stars and Galaxies Form? (Princeton, NJ: Princeton Univ. Press) 67

Marigo, P., Chiosi, C., & Kudritzki, R–P 2003, A&A, 399, 617; see also

Marigo, P., Girardi, L., Chiosi, C., & Wood, P. R. 2001, A&A, 371 152; see also
Murphy, L. J. et al. 2021, MNRAS, 501, 2745

Matthews, T. A., & Sandage, A. R. 1963, ApJ, 138, 30

Miralda-Escude, J. 1999, in The First Stars, Weiss, A., Abel, T. G., & Hill, V. (eds.) (Berlin: Springer–Verlag) 246.

Naidu, R. P., et al. 2022, ApJ, 940, 14

Norman, M. L., Abel, T., & Bryan, G. 1999, in The First Stars, Weiss, A., Abel, T. G., & Hill, V. (eds.) (Berlin: Springer-Verlag) 250

Pan, T., Kasen, D., & Loeb, A. 2012, MNRAS, 422, 2701

Salpeter, E. E. 1964, ApJ, 140, 796

Saslaw, W. C., & Zipoy, D. 1967, Natur, 216, 976

Schaerer, D. 2002, A&A, 382, 28

Shields, G. A. 1999, PASP, 111, 760

The Event Horizon Telescope Collaboration 2019, ApJ, 875, L4

The Event Horizon Telescope Collaboration 2022, ApJ, 930, L12

Weinberg, 1977, The First Three Minutes (New York: Basic Books, Inc.) 74; see also
Loeb, A. 2010, How Did the First Stars and Galaxies Form? (Princeton, NJ: Princeton Univ. Press) 27

Woosley, S. E., Heger, A., & Weaver, T. A. 2002, RvMP, 74, 1015

Zel'dovich, Y. B. 1964, DokAN, 155, 67

Appendix A

Forms of the Hertzsprung–Russell Diagram

As shown in Figure 2.3, the original form of the Hertzsprung–Russell (H–R) diagram, as given by Henry Norris Russell (1877–1957) in 1913 following a similar plot obtained by Ejnar Hertzsprung (1873–1967), was a scatter plot of the absolute magnitudes of stars against their spectral types. In order to obtain the absolute magnitude from the apparent magnitude as measured at the telescope, it is necessary to know the distance to the star (or equivalently its parallax). Because the spectral type is correlated with the surface (or effective) temperature of the star, the plot corresponds to the intrinsic brightness of a star versus its temperature.

After the introduction of photoelectric photometry in the 1950s, this diagram was updated to become a plot of the absolute visual magnitude M_V against a photoelectric color index, which is also correlated with the temperature. The original Johnson broad-band colors—ultraviolet (U), blue (B), and visual (V)—were later expanded into the Johnson–Cousins system with the addition of red (R) and infrared (I) broad-band colors. Most versions of the H–R diagram from that era were plots of M_V versus ($B-V$). For some very hot stars, such as white dwarfs, astronomers chose instead M_V versus ($U-V$), while for very cool stars, such as faint red dwarfs or brown dwarfs, astronomers might use instead M_V versus ($V-I$).

Later, other photometric systems were introduced for specific purposes (Bessell 2005). For example, in 1963, Bengt Strömgren (1908–1987) introduced a four-color narrow-band system consisting of the colors ultraviolet (u), violet (v), blue (b), and yellow (y), which had the advantage of allowing easy determinations of the Balmer discontinuity and a measure of the metallicity in a stellar spectrum. The corresponding version of the H–R diagram is a plot of M_v versus ($b-y$). Similarly, other photometric systems have since been defined, and each has its own corresponding version of the H–R diagram. For example, in the 1990s, the Sloan Digital Sky Survey introduced the ultraviolet (u), green (g), red (r), infrared (i), and far infrared (z) system, and the Wide Field and Planetary Camera 2 (WFPC2) aboard the Hubble Space Telescope adopted the five wide-band filters F336W, F439W, F555W, F675W, and F814W filters, each with their own versions of the H–R diagram.

doi:10.1088/2514-3433/acce33ch23

Another form of the H–R diagram can be used for distant globular star clusters. A typical globular cluster is at least several kiloparsecs away, while the cluster diameter is of the order of 20–100 parsecs. On the scale of the distance to the cluster, all the stars thus lie at roughly the same distance. Consequently, the conversion from apparent magnitude to absolute magnitude is approximately the same for every star in the cluster. Accordingly, a plot of, e.g., the apparent visual magnitudes V of the cluster stars versus their $(B-V)$ colors—the so-called "cluster–magnitude (C–M) diagram"—is essentially the same as the H–R diagram. That is, all the stars would be shifted vertically in magnitude by the same amount.

The final form of the H–R diagram we consider here is what one might call the "theorists' H–R diagram." This is obtained from the familiar M_V versus $(B-V)$ form by first applying the "bolometric correction" (B.C.) to convert the absolute visual magnitude M_V into the absolute bolometric magnitude $M_{bol} = M_V + $ B.C. The bolometric magnitude is a measure of the radiation emitted by the star at all wavelengths. It is thus brighter than M_V, which measures only the flux of radiation emitted at visual wavelengths (i.e., it is less positive because of the peculiar way astronomical magnitudes are determined). Consequently, the B.C. is always negative, and it depends upon the temperature (or color) of the star. It is least negative for the F stars, and it becomes increasingly negative both for increasingly blue and for increasingly red stars. The bolometric magnitude is simply related to the logarithm of the star's luminosity in solar units—$M_{bol} = (5/2) \log (L/L_{Sun}) + 4.72$—which is a typical output of a stellar-evolution calculation. The photometric color index, $(B-V)$, is related to the effective temperature of the star, T_{eff}, which can be obtained, for example, from a model-atmosphere calculation. These two transformations—of M_{bol} to $\log (L/L_{Sun})$ and of $(B-V)$ to T_{eff}—complete the transformation from the observers' H–R diagram, M_V versus $(B-V)$, to the theorists' version, $\log (L/L_{Sun})$ versus T_{eff}.

Reference

For a comprehensive survey of photometric systems, see; Bessell, M. S. 2005, ARA&A, 43, 293

Appendix B

Some Physical Properties of the Sun

For reference, current values for some physical properties of the Sun (Cox 2000a) are listed below:

Mass	$= M_{Sun}$	$= 1.9891 \times 10^{33}$ g	$= 1.9891 \times 10^{30}$ kg
Radius	$= R_{Sun}$	$= 6.95508 \times 10^{10}$ cm	$= 6.995508 \times 10^{5}$ km
Luminosity	$= L_{Sun}$	$= 3.846 \times 10^{33}$ erg s^{-1}	$= 3.846 \times 10^{26}$ watts
Effective temperature	$= T_{eff}$	$= 5777$ K	
Mean density		$= 1.409$ g·cm^{-3}	
Surface gravity	$= g_{Sun}$	$= 2.740 \times 10^{4}$ cm·s^{-2}	
		$= 27.96 \times$ surface gravity at Earth's equator (Cox 2000b)	
Apparent visual magnitude	$= m_V$	$= -26.75$	
Absolute visual magnitude	$= M_V$	$= +4.82$	
Absolute bolometric magnitude	$= M_{bol}$	$= +4.74$	
Spectral type	$= $ G2v		
Age of Sun	$= 4.6 \times 10^{9}$ years		
Central temperature*	$= T_c$	$= 15.48 \times 10^{6}$ K	
Central density*	$= \rho_c$	$= 150.5$ g·cm^{-3}	
Central pressure*	$= P_c$	$= 2.338 \times 10^{17}$ dynes·cm^{-2}	

*Source: BS2005-AGS, OP (Bahcall *et al.* 2005).

References

Bahcall, J. N., Serenelli, A. M., & Basu, S. 2005, ApJ, 621, L85

Cox A. N. (ed) 2000a, Allen's Astrophysical Quantities (New York: AIP/Springer) 12, 340

Cox A. N. (ed) 2000b, Allen's Astrophysical Quantities (New York: AIP/Springer) 245

Appendix C

Zero-Age Main Sequence Stars

This appendix lists a number of physical characteristics from models for zero-age Main Sequence stars as computed using the program ZAMS from Hansen & Kawaler (1994).

Selected Characteristics of Zero-Age Main Sequence Models*

M/M_{Sun}	R/R_{Sun}	$\log(L/L_{Sun})$	T_{eff}	$\log(T_{eff})$	$\log(P_c)$	$\log(T_c)$	$\log(\rho_c)$
0.10	0.116	−3.023	2985	3.475	17.68	6.654	2.736
0.20	0.218	−2.238	3411	3.533	17.24	6.815	2.255
0.30	0.295	−1.957	3451	3.538	17.05	6.880	2.039
0.40	0.380	−1.723	3483	3.542	17.04	6.911	2.017
0.50	0.511	−1.419	3573	3.553	17.10	6.956	2.000
0.60	0.562	−1.172	3926	3.594		6.969	1.898
0.75	0.695	−0.728	4560	3.659		7.031	1.911
0.90	0.849	−0.262	5395	3.732	17.11	7.124	1.895
1.00	0.997	−0.042	5649	3.752	17.17	7.159	1.918
1.50	1.319	0.759	7798	3.892	17.28	7.280	1.885
2.00	1.481	1.262	9817	3.992	17.21	7.324	1.672
3.00	1.835	1.951	13,122	4.118	17.06	7.371	1.553
5.00	2.470	2.773	18,155	4.259	16.84	7.422	1.279
10.00	3.730	3.772	26,242	4.419	16.57	7.484	0.921
20.00	5.569	4.631	35,237	4.547	16.37	7.535	0.624
30.00	6.978	5.066	40,365	4.606	16.29	7.560	0.484
40.00	8.180	5.345	43,853	4.642	16.26	7.575	0.396
60.00	10.203	5.701	48,195	4.683	16.22	7.594	0.286

*From Hansen & Kawaler (1994, pp. 45, 46).

doi:10.1088/2514-3433/acce33ch25

Reference

Hansen, C. J., & Kawaler, S. D. 1994, Stellar Interiors (New York: Springer)

Appendix D

Electron Degeneracy

When an astrophysicist speaks of electrons being "degenerate," what she means is that the temperature is close enough to absolute zero so that electrons fill up all the lowest energy states available. This has been known to occur since 1925, when physicist Wolfgang Pauli (1900–1958) articulated the "Pauli exclusion principle," which states that no two electrons with exactly the same quantum numbers can occupy the same energy state. This principle explains the basic structure of many-electron atoms. In such an atom, the main energy levels are characterized by the principal (integer) quantum number, n. An electron in a level with a given value of n also has angular momentum, the magnitude of which is characterized by the integer quantum number l. In quantum mechanics, it is also possible to measure one component of the vector angular momentum—usually termed the z-component, where z refers to, e.g., the axis of rotation of the atom—and the quantum number characterizing this quantity is usually denoted by the integer m. In addition, an elementary particle like an electron has an intrinsic spin angular momentum, usually denoted by s. An electron has two possible spin states, generally called "up" or "down," and the corresponding spin quantum numbers are $s = \frac{1}{2}$ or $s = -\frac{1}{2}$. Thus an electron in an atomic energy state is uniquely characterized by the four quantum numbers n, l, m, and s.

In ordinary atomic hydrogen, which contains one electron and one proton, the ground state is $n = 1$; the only allowed values for the angular momentum quantum numbers are $m = l = 0$; and the spin s can be either up or down. Atomic helium has two electrons in orbit around a nucleus containing two protons (and one or more electrically neutral neutrons, which essentially do not interact with the electrons). Again the ground state has $(n,l,m) = (1,0,0)$, so this state can accommodate two electrons, one with spin up, and the other with spin down. The next lightest element, lithium (Li), has three electrons around a nucleus containing three protons (and different numbers of neutrons in different isotopes). Again, the ground state has the quantum numbers $(n,l,m) = (1,0,0)$, so according to the Pauli principle it can only

doi:10.1088/2514-3433/acce33ch26

hold two of the three electrons. The third electron has to go into the next highest energy level, with $n = 2$. In this level, the angular-momentum quantum number l can take on the values $l = 1$ or $l = 0$, and in the state with $l = 1$, the quantum number m can take on the values, $m = 1$, $m = 0$, or $m = -1$. Thus, there are four distinguishable states with different values of l and m—$(l,m) = (1,1)$, $(1,0)$, $(1,-1)$, and $(0,0)$—and since each state can hold electrons with spin up and spin down, the energy level $n = 2$ can accommodate up to 8 electrons. Atoms with filled shells—like He with atomic number 2, referring to the two electrons, and Ne, with atomic number 10, corresponding to the two electrons that fill the first energy level and the eight more that fill the second—are the noble gases. Continuing in this way, one can fill out the energy-level structures of the atoms in the rest of the periodic table of elements.

The same type of argument applies to a gas of free electrons. For the moment, let us ignore the charge of the electron, so that we do not have to worry about electrostatic interactions, and let us suppose that the electrons are confined within a one-dimensional box of side length L. Then, the Heisenberg uncertainty principle tells us that there is a minimum value of the momentum for the particle in that box: $p = h/L$, where h is Planck's constant. Just as in the case of the hydrogen atom, this lowest energy level can hold exactly two electrons (one with spin up and one with spin down). The next energy level in this one-dimensional example can hold two more electrons, and so on. We can extend this reasoning to three dimensions by considering the "box" to have dimensions in each of the three spatial directions x, y, and z. If we have a large number N of electrons to put into the box (corresponding to increasing the electron number density, if the box has a fixed size), then it is clear that the electrons will continue to fill all the available energy levels up to a finite value called the Fermi energy—after the great Italian physicist Enrico Fermi (1901–1954)—which can be calculated from the electron density at zero temperature. Such a zero-temperature gas of free electrons is said to be degenerate. Even at zero temperature, a degenerate electron gas exerts a finite pressure because the individual electrons have finite energies and momenta.

At finite temperatures, some of the electrons with kinetic energies near the Fermi level at the top of the energy distribution can be excited to higher, unoccupied energy states, just as electrons in atoms can be excited to unoccupied higher levels. If the characteristic thermal energy of an electron gas is small compared to the Fermi energy, the properties are not very different from those of a fully degenerate electron gas. Because the densities inside a brown dwarf or white dwarf star are so high, the electrons are in effect "squeezed" out of the atoms, leaving a sea of degenerate electrons with embedded atomic nuclei to preserve overall charge neutrality. This is the state of matter inside most of a white dwarf or brown dwarf.

Because neutrons are spin-½ particles like electrons (generically termed "fermions"), similar ideas hold for the sea of degenerate neutrons inside a neutron star. However, although neutrons do not interact with other particles through electrostatic (Coulomb) interactions, they do interact via the strong nuclear force that is responsible for binding neutrons and protons (collectively termed "nucleons") together inside an atomic nucleus. The strong force cannot be ignored as easily as the electrostatic force in calculating the energy levels of matter inside a neutron star. Thus, although the general concept of degenerate neutrons holds, the physical properties of the matter inside a neutron star depend much more on the interactions among the particles than is the case for the matter in a white dwarf.

Appendix E

Properties of Some Brown Dwarf Stars

The table below summarizes results for a selected few objects spanning the three brown-dwarf spectral classes. The spectral type is listed in the column headed Sp; T_{eff} is the effective temperature in Kelvins; $\log g$ is the logarithm of the surface gravity in cm s^{-2}; $\log L$ is the logarithm of the luminosity in units of the solar luminosity; R and M are, respectively, the radius and mass in units of Jupiter's radius and mass; D is the distance to the object in parsecs; and the final column gives the sources for the data, which are listed below the table. In some cases, the references give a range of values for the different quantities; for them, the data listed in the table are simple averages of the extremes. Readers interested in the actual values should consult the original sources.

Name	Sp	T_{eff}	$\log g$	$\log L$	R	M	D	References
GD165B	L3	1750	5.00	−4.06	0.99	76	32 pc	(1, 2)
Gl 229B	T7pec	927±77	4.75	−5.21	1.26	70±5	6 pc	(2, 3, 4, 5, 6)
2MASSI J1217−0311	T7.5	885±75	5.11	−5.31	0.89	45		(6, 7, 8)
HD 3651B	T7.5	810	5.30	−5.60	0.78	56		(7, 8, 9, 10)
Gl 570D	T7.5	759±63	5.16	−5.54	0.84	39		(4, 6, 7, 8, 9)
2MASSI J0415−0935	T8	677±56	5.20	−5.67	0.84	45		(6, 7, 8, 9)
UGPS 0722−05	T9	569±45	4.50	−6.03	1.18	8	11 pc	(6, 11)
WISEPC J0148−7202	T9.5	526±88	4.63	−5.24	1.00	17	14 pc	(6, 11)
WISEPC J2056+1459	Y0	464±88	4.75	−5.79	0.94	21	4 pc	(6, 11)
WISEPA J0410+1502	Y0	451±88	4.00	−5.29	1.16	6	12 pc	(6, 11)
WISEPA J1738+2733	Y0	450±88	4.88	−5.83	0.90	25	5 pc	(6, 11)
WISEPA J1541−2250	Y1	395±88	4.38	−5.82	1.04	10	9 pc	(6, 11)

References.
(1) Kirkpatrick *et al.* (1999), (2) Leggett *et al.* (2002), (3) Saumon *et al.* (2000), (4) Saumon *et al.* (2006), (5) Nakajima *et al.* (2015), (6) Kirkpatrick *et al.* (2021), (7) Knapp *et al.* (2004), (8) Saumon *et al.* (2007), (9) Leggett *et al.* (2010), (10) Faherty *et al.* (2010), (11) Cushing *et al.* (2011).

References

Cushing, M. C., et al. 2011, ApJ, 743, 50

Faherty, J. K., et al. 2010, AJ, 139, 176

Kirkpatrick, J. D., et al. 1999, ApJ, 519, 802

Kirkpatrick, J. D., et al. 2021, ApJS, 253, 7

Knapp, G. R., et al. 2004, AJ, 127, 3553

Leggett, S. K., et al. 2002, ApJ, 564, 452

Leggett, S. K., et al. 2010, ApJ, 710, 1627

Nakajima, T., Tsuji, T., & Takeda, Y. 2015, AJ, 150, 53

Saumon, D., et al. 2000, ApJ, 541, 374

Saumon, D., et al. 2006, ApJ, 647, 552

Saumon, D., et al. 2007, ApJ, 656, 1136

Appendix F

The Laser Interferometer Gravitational-wave Observatory (LIGO)[1]

Albert Einstein's (1879–1955) general theory of relativity, published in 1916, predicted the existence of gravitational waves. Because they are so weak, technology capable of detecting them directly did not exist in the 20th century. Nevertheless, in the 1960s, American physicist Joseph Weber (1919–2000) pioneered the use of massive resonant bars as potential gravitational-wave detectors, and in 1962, two Soviet physicists proposed the use of optical interferometry[2] for this purpose (Gertsenshtein & Pustovoit 1962). In the late 1960s, American physicists Rainer Weiss at MIT (Weiss 1972) and Robert L. Forward (1932–2002) at the Hughes Research Laboratories began the construction of prototype interferometric detectors. The use of free-swinging mirrors in such gravitational-wave detectors was initiated by Weiss and by physicists in Germany and Scotland in the 1970s.

In 1980, the National Science Foundation (NSF) provided funding to MIT for the construction of a large interferometer, which established the feasibility of using kilometer-scale interferometers to detect gravitational waves. At the same time, similar work was being undertaken at Caltech, and because of the costliness of these projects, NSF forced them to combine. The project struggled with management issues until the early 1990s, when Caltech's Barry C. Barish—who had established a reputation as an excellent manager of large particle-physics projects—was named the project manager. Under Barish's leadership, the project developed a proposal to construct the Laser Interferometer Gravitational-wave Observatory (LIGO) in

---

[1] en.wikipedia.org/wiki/LIGO.

[2] In optical interferometry, two coherent beams of light (e.g., from a laser) are caused to coincide at a detector. Because of the wave nature of light, the two beams interfere with each other, producing a local peak of intensity when the two beams are in phase with each other (so that the peaks and troughs of the original beams coincide: "constructive interference") and producing a local trough when they are completely out of phase (so that a peak of one beam coincides with a trough in the other: "destructive interference"). This phenomenon was first observed by the English scientist Thomas Young in 1800. The resulting interference pattern is extremely sensitive to the difference in the distances traveled by the two beams.

stages, with successive improvements to reduce the noise due to ambient sources—such as seismic vibrations—enough to allow detection of the very faint gravitational-wave signals. The proposal was approved by NSF in 1994 at the level of $395 million, the largest amount the agency had ever supported, and construction began that same year.

As shown in Fig. 19.1, LIGO consists of two essentially identical L-shaped interferometers—each with two highly evacuated, 4-km-long arms (about 2.5 miles), one located in Hanford, Washington, and the other in Livingston, Louisiana. Large, suspended mirrors are located at the far ends of each arm. The interferometric detectors are so sensitive that they are able to measure changes caused by the passage of a gravitational wave that are smaller than one ten-thousandth of the diameter of an atomic nucleus in the length of an arm of the interferometer. The original phase of the project was completed and began taking data in 2002, but the initial years of operation were devoted primarily to identifying and eliminating or compensating for sources of noise in the system. After two significant upgrades, the improved detectors began taking data in 2015, and the first actual detection of gravitational waves occurred on 2015 September 14. Detected by both LIGO observatories and the French–Italian Virgo detector, this event—labelled GW150914—proved to be gravitational waves emitted during the inspiral and coalescence of two ~30 M_{Sun} black holes (Abbott *et al.* 2016). For this discovery, Barry Barish, Caltech theorist Kip S. Thorne—who had been investigating gravitational-wave emission from various sources since the 1960s—and Rainer Weiss were awarded the Nobel Prize in 2017.

By the end of 2019, LIGO had completed three runs of data taking, which resulted in the detection of some 50 sources of gravitational waves. Operations were suspended in March 2020 because of the COVID-19 pandemic. As of this writing, LIGO's fourth observing run—in coordination with Virgo and the Japanese gravitational-wave observatory KAGRA—is scheduled to begin in mid-December 2022.

References

Abbott, B. P., et al. 2016, PhRvL, 115, 061102

Gertsenshtein, M. E., & Pustovoit, V. J. 1962, SvPh-JETP, 43, 605

Weiss, R. 1972, MEQPR, 105, 54; see references in https://en.wikipedia.org/wiki/Rainer_Weiss

Appendix G

Zero-Age Main Sequence Models for the First Stars

This Appendix lists a number of physical characteristics from models for zero-age Main Sequence stars from Baraffe *et al.* (2001) and Schaerer (2002). The expression for the Eddington luminosity, L_{Edd}, is taken from Hansen and Kawaler 1994, p. 47.

Selected Characteristics for Zero-Age Main Sequence Models for the First Stars*

M/M_{Sun}	$\log(L/L_{Sun})$	L/L_{Edd}	T_{eff}	$\log(T_{eff})$	R/R_{Sun}	$\log(T_c)$
5	2.870	0.00	27,540	4.440	1.2	
9	3.709	0.02	41,880	4.622	1.4	
15	4.324	0.04	57,410	4.759	1.5	
25	4.890	0.09	70,800	4.850	1.9	
40	5.420	0.19	79,430	4.900	2.7	
60	5.715	0.25	87,700	4.943	3.1	
80	5.947	0.32	93,330	4.970	3.6	
120	6.243	0.42	95,720	4.981	4.8	8.140
200	6.574	0.54	99,770	4.999	6.5	8.155
300	6.819	0.63	101,620	5.007	8.3	8.167
400	6.984	0.70	106,660	5.028	9.1	
500	7.106	0.74	106,900	5.029	10.4	8.179
1000	7.444	0.80	106,170	5.026	15.6	

*From Baraffe *et al.* (2001) and Schaerer (2002).

References

Baraffe, I., Heger, A., & Woosley, S. E. 2001, ApJ, 550, 890
Hansen, C. J., and Kawaler, S. D. 1994, Stellar Interiors (New York: Springer-Verlag), 47
Schaerer, D. 2002, A&A, 382, 28

Bibliography

Abell, G. O., Morrison, D., & Wolff, S. C. 1987, Exploration of the Universe (Philadelphia, PA: Saunders College Publishing)

Aitken, R. G. 1918, The Binary Stars (New York: D. C. McMurtrie)

Alimi, J. M. & Fuzfa A. (ed) 2006, Proc. Albert Einstein Century Int. Conf. AIP Conf. Proc. Vol. 861 (Melville, NY: AIP Publishing)

Allen, C. W. 1955, Astrophysical Quantities 2nd ed (Univ. London: Athlone Press)

Aller, L. H., & McLaughlin, D. B. 1965, Stellar Strcture: Stars and Stellar Systems Vol. 8 (Chicago, IL: Univ. Chicago Press)

Aspray, W. 1986, An Interview with Martin Schwarzschild, OH 124 (Minneapolis, MN: Charles Babbage Institute, Univ. Minnesota)

Baschek, B. Kegel, W. H. & Traving G. (ed) 1975, Problems in Stellar Atmospheres and Envelopes (New York: Springer)

Battaner, E. 1996, Astrophysical Fluid Dynamics (Cambridge: Cambridge Univ. Press)

Battrick B. (ed) 1997, Hipparcos Venice'97 (Noordwijk: ESA)

Berthomieu, G. & Cribier M. (ed) 1990, Inside the Sun (Dordrecht: Kluwer)

Bethe, H. A. 1991, The Road from Los Alamos (New York: AIP)

Branch, D., & Wheeler, J. C. 2017, Supernova Explosions (Berlin: Springer Nature)

Brown T. M. (ed) 1993, Gong 1992: Seismic Investigations of the Sun and Stars ASP Conf. Ser. Vol. 42 (San Francisco, CA: ASP)

Burke, B. F., & Graham-Smith, F. 1997, An Introduction to Radio Astronomy (Cambridge: Cambridge Univ. Press)

Cassisi S. (ed) 2005, Evolution of Stars and Stellar Populations (New York: Wiley)

Catelan, M., & Smith, H. A. 2015, Pulsating Stars (Weinheim: Wiley-VCH GmbH)

Chandrasekhar, S. 1939, An Introduction to the Study of Stellar Structure (Chicago, IL: Univ. Chicago Press) reprinted 1957 (New York: Dover)

Clayton, D. D. 1968, Principles of Stellar Evolution and Nucleosynthesis (New York: McGraw-Hill)

Cox A. N. (ed) 2000, Allen's Astrophysical Quantities (New York: AIP/Springer)

Cox, J. P. 1980, Theory of Stellar Pulsation (Princeton, NJ: Princeton Univ. Press)

Cox, J. P. & Hansen C. J. (ed) 1982, Pulsations in Classical and Cataclysmic Variable Stars (Boulder, CO: JILA and NBS)

Dalrymple, B. G. 2004, Ancient Earth, Ancient Skies: The Age of the Earth and Its Cosmic Surroundings (Stanford, CA: Stanford Univ. Press)

Eddington, A. S. 1926, The Internal Constitution of the Stars (Cambridge: Cambridge Univ. Press)

Eppley, K. R. 1975, The Numerical Evolution of the Collision of Two Black Holes, (Princeton Univ.) PhD thesis

Gamow, G. 1945, The Birth and Death of the Sun (New York: Penguin Books)

Greenstein J. L. (ed) 1960, Stellar Atmospheres (Chicago, IL: Univ. Chicago Press)

Halpern, P. 2021, Flashes of Creation (New York: Basic Books)

Hansen, C. J., & Kawaler, S. D. 1994, Stellar Interiors (New York: Springer)

Hawking, S. 1988, A Brief History of Time (New York: Bantam Books)

Holberg, J. B. 2007, Sirius: Brightest Diamond in the Night Sky (Chichester: Praxis Publishing Company)

Jones, H. R. A. & Steele I. A. (ed) 2001, Ultracool Dwarfs (Heidelberg: Springer)

Jones-Krull, C. M. Browning, M. K. & West A. A. 2011, 16th Cambridge Workshop on Cool Stars, Stellar Systems, and the Sun, ASP Conf. Ser. Vol. 448 (San Francisco, CA: ASP)

Kafatos, M. C., Harrington, R. S., & Maran, S. P. 1986, Astrophysics of Brown Dwarfs (Cambridge: Cambridge Univ. Press)

Kaplan, I. 1955, Nuclear Physics (Reading, MA: Addison-Wesley)

Kellerman, K., & Sheets, B. 1983, Serendipitous Discoveries in Radio Astronomy (Green Bank, WV: National Radio Astronomy Observatory)

Kippenhahn, R., & Weigert, A. 1990, Stellar Structure and Evolution (Berlin: Springer)

Kukarkin, B. V., et al. 1969, General Catalogue of Variable Stars, Volume 1: Constellations Andromeda—Grus (Moscow: Astronomical Council of the Academy of Sciences of the USSR: Sternberg Institute of the Moscow State Univ.)

Langer, W. L. 1968, An Encyclopedia of World History (Boston, MA: Houghton Mifflin Company)

Ledoux, P. Noels, A. & Rodgers A. W. (ed) 1974, Stellar Instability and Evolution, Proc. IAU Symp. # 59 (Switzerland: Springer Nature)

Lequex, J. 2013, Birth, Evolution and Death of Stars (Singapore: World Scientific)

Loeb, A. 2010, How Did the First Stars and Galaxies Form? (Princeton, NJ: Princeton Univ. Press)

Longair, M. S. 2008, Galaxy Formation 2nd ed (New York: Springer)

Luyten W. J. 1971, White Dwarfs, Proc. IAU Symp. # 42 (Dordrecht: Springer)

Manchester, R. N., & Taylor, J. H. 1977, Pulsars (San Francisco, CA: Freeman)

Mannings, V. Boss, A. P. & Russell S. S. (ed) 2000, Protostars and Planets IV (Tucson, AZ: Univ. Arizona Press)

Mihalas, D. 1970, Stellar Atmospheres (San Francisco, CA: Freeman)

Moore, D. 2020, What Stars Are Made Of: The Life of Cecilia Payne-Gaposchkin (Cambridge, MA: Harvard Univ. Press)

Neugebauer, G., & Leighton, R. B. 1969, Two-Micron Sky Survey: A Preliminary Catalog, NASA SP-3047

Osaki, Y. & Shibahashi H. (ed) 1970, Progress of Seismology of the Sun and Stars (Berlin: Springer)

Pagel, B. E. J. 1997, Nucleosynthesis and Chemical Evolution of Galaxies (New York: Cambridge Univ. Press)

Pannekoek, A. 1961, A History of Astronomy (London: George Allen & Unwin Ltd.) reprinted 1989 (New York: Dover)

Payne, C. H. 1925, Stellar Atmospheres: A Contribution to the Observational Study of High Temperatures in the Reversing Layers of Stars, (Radcliffe College) PhD thesis

Payne-Gaposchkin, C. 1957, The Galactic Novae (New York: Interscience)

Peimbert, M. & Jugaku J. 1987, Star Forming Regions, Proc. IAU Symp. #115 (Dordrecht: Reidel)

Philip, A. G. D. Hayes, D. & Leibert J. 1987, Second Conf. on Faint Blue Stars, Proc. IAU Colloq. #95 (Schenectady, NY: L. Davis Press)

Preston, M. A. 1962, Physics of the Nucleus (Reading, MA: Addison-Wesley)

Prialnik, D. 2000, An Introduction to the Theory of Stellar Structure and Evolution (Cambridge: Cambridge Univ. Press)

Provost, J., & Schmider F. X. (eds) 1997, Sounding Solar and Stellar Interiors (Dordrecht: Springer)

Rasio, F. A., & Stairs I. H. 2005, Binary Radio Pulsars, ASP Conf. Ser. Vol. 328 (San Francisco, CA: Astronomical Society of the Pacific)

Rebolo, R. Martin, E. & Zapatero Osario M. R. (eds) 1998, Brown Dwarfs and Extrasolar Planets, ASP Conf. Ser. Vol. 134 (San Francisco, CA: ASP)

Reid, I. N., & Hawley, S. L. 2005, New Light on Dark Stars (Chichester: Springer/Praxis Publishing)

Sagan, C. 1980, Cosmos (New York: Random House)

Schatzman, E. 1958, White Dwarfs (Amsterdam: North Holland)

Schwarzschild, M. 1958, Structure and Evolution of the Stars (Princeton, NJ: Princeton Univ. Press)

Shapiro, S. L., & Teukolsky, S. A. 1983, Black Holes, White Dwarfs, and Neutron Stars (New York: Wiley)

Smarr, L. L. 1975, The Structure of General Relativity with a Numerical Illustration: The Collisioni of Two Black Holes, (Univ. Texas) PhD thesis

Sobel, D. 2016, The Glass Universe: How the Ladies of the Harvard Observatory Took the Measure of the Stars (New York: Viking Press)

Tarter, J. C. 1975, The Interactions of Gas and Galaxies within Galaxy Clusters, (Univ. California) PhD thesis

Thomas R. N. (ed) 1961, Aerodynamic Phenomena in Stellar Atmospheres Proc. 4th Symp. on Cosmical Gas Dynamics (Bologna: Nicola Zanichelli)

Ulrich, R. K., Rhodes, E. J., & Däppen W. (eds) 1995, GONG'94: Helio- and Astero-Sesimology ASP Conf. Ser. Vol. 76 (San Francisco, CA: ASP)

Unno, W., Osaki, Y., Ando, H., & Shibahashi, H. 1979, Nonradial Oscillations of Stars (Tokyo: Univ. Tokyo Press)

Van Horn, H. M. & Weidemann V. (eds) 1979, Proc. IAU Colloq. 53 White Dwarfs and Variable Degenerate Stars (Rochester, NY: Univ. Rochester Press)

Warner, B. 2003, Cataclysmic Variable Stars (Cambridge: Cambridge Univ. Press)

Weinberg, S. 1977, The First Three Minutes (New York: Basic Books, Inc) updated 1988

Weiss, A. Abel, T. G. & Hill V. 1999, The First Stars, (Berlin: Springer) Proc. MPA/ESO Workshop Held (Garching, Germany, 4–6 August 1999)

Weltevrede, P. Perera, B. B. P. Preston, L. L. & Sanidas S. (ed) 2018, Pulsar Astrophysics—The Next 50 Years Proc. IAU Symp. # 337 (Cambridge: Cambridge Univ. Press)

Witten L. (ed) 1962, Gravitation: An Introduction to Current Research (New York: Wiley)

Yeager, A. J. 2021, Bright Galaxies, Dark Matter, and Beyond: The Life of Astronomer Vera Rubin (Cambridge, MA: MIT Press)

Printed in the USA
CPSIA information can be obtained
at www.ICGtesting.com
LVHW071042291023
762114LV00015B/26